This book is to be returned on
or before the date stamped below

AEROSPACE STRATEGIC TRADE

Aerospace Strategic Trade

How the U.S. subsidizes the large commercial aircraft industry

PHILIP K. LAWRENCE
Director Aerospace Research Centre,
UWE Bristol
and
Director CERMAS ESC, Toulouse

Ashgate

Aldershot • Burlington USA • Singapore • Sydney

Published by
Ashgate Publishing Ltd
Gower House
Croft Road
Aldershot
Hants GU11 3HR
England

Ashgate Publishing Company
131 Main Street
Burlington, VT 05401-5600 USA

Ashgate website: http://www.ashgate.com

British Library Cataloguing in Publication Data
Lawrence, Philip K., 1953-
    Aerospace strategic trade : How the U.S. subsidizes the
    Americam large commercial aircraft industry
    1.Aircraft industry - Research grants - United States
    I.Title
    338.4'36291334'0973

Library of Congress Control Number: 00-110692

ISBN 0 7546 1696 7

Printed and bound in Great Britain by MPG Books Ltd, Bodmin, Cornwall

Contents

Preface

Please remember that the United States did not build the first high performance fighters or the first jet engine or the first ballistic rocket or the first commercial jet aircraft. However, over the last 80 years whenever the federal government stepped up and established a national priority supported by adequate funding, we moved forward to achieve dominance, (Senate Armed Services Committee: Subcommittee on Acquisition and Technology, April 10 1997).

Overview

According to the orthodox view the U.S. economy is run on free market or laissez faire principles, which means that U.S. policy makers do not provide government support for industrial or commercial sectors. In many instances this is doubtless true, but it is not the case with strategic industries, such as aerospace. Support for the aerospace sector has been viewed as essential, because aerospace technologies have been the material backbone of U.S. security systems, built around the centrality of the U.S. Air Force in defence planning. During the Cold War era it was inconceivable that the United States could leave the fate of military aerospace to the free market, as they faced an adversary capable of destroying their cities and infrastructure.

There is nothing remotely contentious about the U.S. implementing an industrial policy for its defence and aerospace industries. Indeed, the West's security during the Cold War depended on such a policy. But what must be realized is that America's historic dominance in commercial aerospace, and particularly the large commercial aircraft sector, arose on the back of defence technology paid for by the U.S. government. As leading U.S. industrial economists Mowery and Rosenberg note '... the history of technical development in commercial aircraft consists largely of the utilization for commercial purposes of technical knowledge developed for military programs at government expense', (1982, p.140).

This study examines U.S. federal financial support for the American Large Commercial Aircraft (LCA) Industry during the years 1992-1998. The aim of the work is to show how the U.S. conducts strategic trade in aerospace via an industrial policy based on R&D supports. The analysis utilizes a database on U.S. government expenditure for aerospace, compiled between 1996 and 1998. However, the analysis based on these data is supplemented with data from NASA Research & Technology (R&T) contracts, particularly for the years 1996 and 1997. Accordingly, two kinds of data are provided here. For the years 1992-98 we give data on the flow of U.S. government funds into the aerospace industry and its subcomponent the large commercial aircraft sector. This information comes from detailed budget titles obtained from the U.S. Congress and other sources listed in Appendix B. However, in addition for 1996/97 there is data from NASA R&D contracts that shows a flow of benefits to Boeing and McDonnell Douglas LCA activities. The former is characterized as a "top down" approach, the latter as "bottom up".

The implicit frame of reference for this study is the World Trade Organization (WTO) rules and disciplines on subsidy, which are to be found in the WTO Agreement on Subsidies and Countervailing Measures, (ASCM).

Our analysis of federal policy for large commercial aircraft sector (LCA) shows clearly that the U.S. provides substantial financial subsidy to its large commercial aircraft industry. This subsidy comes predominantly from NASA and DoD programs.

NASA Subsidy

Table P.1   NASA Aeronautical Research and Technology Program Budget

(Numbers in U.S. $ mn)

	FY93	FY94	FY95	FY96	FY97	FY98	Total
R&T Base	436.5	448.3	366.3	354.7	404.2	418.3	2428.3
AST	12.4	101.3	150.1	169.8	173.6	211.1	818,3
HSR	117.0	187.2	221.3	233.3	243.1	245.0	1246.9
Other	299.7	83.9	144.3	159.5	23.3	45.7	756.4
TOTAL	865.6	1,020.7	882.0	917.3	844.2	920.1	5250.9

Table P.1 above shows the size of NASA Aeronautical Research and Technology Programs. In the years 1993 – 1998 these budgets had a cumulative value of U.S.$5.25bn.

NASA supports and subsidizes the U.S. large commercial aircraft sector by technology transfer from these publicly funded R&T programs, many of which are undertaken directly by the industry. On completion of the contracts the knowledge and knowhow produced becomes the property of the companies.

NASA funding for R&T is necessary because the American authorities recognize that their private companies are not able or willing to provide sufficient funding for R&T. Historically the funding has been justified by the recognition that the U.S. LCA sector has been falling behind Airbus Industrie in critical areas of advanced aircraft technology.

Figure P.1   Value and Flow of NASA Funding into U.S. LCA Sector

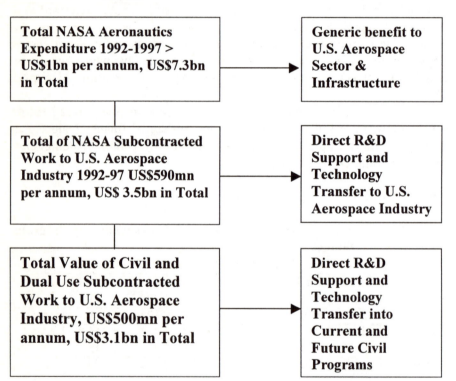

In total we estimate that NASA aeronautics expenditure in the years 1992-1997 was $7.3bn. More pertinently, the value of subcontracted work to the U.S. aerospace industry (R&T contracts awarded to companies) was some $3.5bn, of which $3.1bn had clear civil or dual-use applicability, (see fig. P.1 above).

Department of Defense Subsidy to the LCA Sector

As we have already noted U.S. dominance in commercial aeronautics was built on the bedrock of DoD funding for military aerospace. In the early days of the civil jet business complete airframes and systems were spun-off to commercial applications. Alternatively, as was the case with the B-707, civil and military variants were developed simultaneously to attenuate risk. Today the process of spin-off is more complex and many U.S. sources deny that any military/civil synergies still exist. Our analysis shows this to be completely false. In avionics, design tools, manufacturing technology and materials there are still fundamental synergies, where military funded technologies have what we call spin-off potential. Thus, as we show in chapter 5, Boeing's work on the B2 program enabled it to develop the capability to manufacture large aero-structures made of polymer matrix composites (PMCs). These composite structures continue to play a vital role in the drive to reduce aircraft weight and hence operating costs, but they are notoriously expensive to manufacture. Thus proving the technology at DoD expense has been highly beneficial to Boeing. More recently, as Mr. Phil Condit confirmed in the *Sunday Times,* design tools developed for the JSF program will be used on future Boeing civil programs, (Sunday Times, 12/6/00).

Quantifying DoD Support

All told the sums of money spent by the DoD on aeronautical R&D are far larger than that allocated to NASA. However, it is a complex matter to track precisely the portion of the several billions of dollars spent annually on aircraft related R&D, which supports and subsidizes U.S. large commercial aircraft manufacturing.

In order to estimate the value of DoD subsidy to commercial programs we isolated development and R&T program areas, which had a clear dual-use potential, i.e., a potential for civil applications. We then determined the percentage of contracts in these dual-use programs going to the then two U.S. LCA primes. On that basis we estimate that DoD subsidy to the U.S. LCA sector was approximately $560mn per annum in the years 1992-1997, or 3.4 billion dollars in total. This is shown in table P.2.

Table P.2   **Financial Benefits to U.S. LCA Primes from DOD R&T, Development and Procurement, (US$mn)**

Year	From R&T	From Development	From Procurement	Total
1992	219	219	126	564
1993	253	257	80	590
1994	236	154	70	459
1995	277	213	66	556
1996	272	263	58	594
1997	279	255	60	593
Total	1536	1361	460	3356

Our figures for DoD subsidy include Independent Research and Development (IR&D) funding, which is paid to U.S. companies as an overhead on procurement contracts. This is a very direct form of benefit, which in the past may have represented as much as five percent of an LCA firm's R&D expenditure, (COTA, 1991).

Conclusions

The U.S. Federal Government subsidises its large commercial aircraft (LCA) industry via NASA R&T and DoD R&D programs, which provide technology transfer to the LCA sector. The subsidy derives from:

➤ The NASA R&T Base Program
➤ NASA Aeronautical Focused Programs
➤ DoD Dual-Use R&T and development contracted to LCA primes.

Our analysis indicates that the subsidy coming from these programs amounted to roughly one billion dollars per annum for 1996 and 1997.

Understating Subsidy

The figure given above for U.S. LCA subsidy is undoubtedly conservative for a number of reasons. First, the real value of the technology transferred into LCA programs may be much higher than the actual financial contract values which we detail. Such technology may also yield benefits in the future, in some instances over the whole life of a program. In effect, a company like Boeing is benefiting from decades of government investment in facilities, capital equipment and personnel, which gives the corporation access to unique and priceless competencies provided at government expenses. As developing countries have found to their cost the sunk costs necessary to establish an aerospace industrial base are enormous. But in the case of the U.S. they could be justified because of the Cold War confrontation with the USSR.

A second factor concerns DoD top secret or black programs. Because we could not access the information we have assumed no transfer of benefits from these programs, yet anecdotal evidence shows that engineers working on black programs are often able to use DoD R&T on civil work. Finally, we also assumed no subsidy coming from NASA's enormous space budget. But again we know that technology such as the titanium alloys (ti-metal) used on the B-777's nacelles was developed on the X30 NASP program funded by NASA.

Because of the points cited above we believe that the real benefits going to Boeing from govermnment funding could be more than twice as high as the one billion dollars we cite.

Acknowledgements

In order to complete our analysis we have benefited significantly from the work of Jean-Pierre Picchiottino, who has assembled an extensive database of information on U.S. aerospace funding and details on NASA and DoD R&D contracts, under the provision of the U.S. Freedom of Information Act. We wish to express our thanks for his invaluable input.

Numerous associates from the aerospace industry gave us assistance with this project and we would like to give mention to Dr Tony Parry and Dr Hans-Henrich Altfeld. In addition our colleagues Professor Dieter Schmitt at the University of Munich and Dr David Thornton, of Campbell University, North Carolina are owed thanks for their helpful contributions. Our technology specialist, Dr Anders Hansson, also provided key insights into the technology transfer issue.

Philip Lawrence
Derek Braddon

Aerospace Research Centre
UWE, Bristol 1999

List of Abbreviations

ACP	Advanced Composites for Propulsion Program
ACT	Advanced Composite Technology Program
AECMA	Association of European Aerospace Industries
AEREA	Association of European Research Establishments in Aerospace
AFP	Aeronautics Focused Program
AI	Airbus Industrie
AIA	Aerospace Industries Association (USA)
AI-EU	Aerospace Industry in the EU
AIMS	Aircraft Information Management System
AI-U.S.	Aerospace Industry of the U.S.
ART	Aeronautical Research and Technology Program
ASCM	Agreement on Subsidies and Countervailing Measures
ASI	Agenzio Spaziale Italiana
AST	Advanced Subsonic Technology Program
AT&T	American Telephone and Telegraph
ATP	Advanced Technology Program
AWACS	Airborne Warning and Control System
BE	Belgium
BMDO	Ballistic Missile Defense Office
BNSC	British National Space Council
CAD	Computer Aided Design
CAM	Computer Aided Manufacture
CATIA	Computer-aided, Three-dimensional, Interactive Application
CDTI	Centro Para el Desarrollo Technológico Industrial
CEO	Chief Executive Officer
CFD	Computational Fluid Dynamics
CIRA	Centro Italiano Ricerche Aerospaziali
CNES	Centre National d'Etudes Spatiales
COTA	Congressional Office of Technology Assessment
CRADA	Co-operative Research and Development Agreement
DARA	Deutsche Agentur für Raumfahrtangelegenheiten

DARPA	Defense Advanced Research Project Agency (USA)
DE	Germany
DERA	Defence Evaluation and Research Agency (UK)
DISCS	Domestic International Sales Corporations
DLR	Deutsche Forschungsanstalt für Luft- und Raumfahrt
DNW	Duits Nederlandse Windtunnel
DOC	Department of Commerce (USA)
DOCs	Direct Operating Costs
DoD	Department of Defense (USA)
DOE	Department of Environment (USA)
DOT	Department of Transportation (USA)
DSB	Dispute Settlement Body (WTO)
DSU	Dispute Settlement Understanding (WTO)
ES	Spain
ESA	European Space Agency
ETW	European Transonic Windtunnel
EU	European Union
EUMETS	Europe's Meteorological Satellite Organization
FAA	Federal Aviation Administration (USA)
FBL	Fly- by- light
FEDD	For Early Domestic Distribution
FFA	Aeronautical Research Institute of Sweden
FLASH	Fly by Light Advanced Systems Hardware
FOIA	Freedom of Information Act
FR	France
FSCS	Foreign Sales Corporations
FTE	Full Time Equivalent
FY	Fiscal Year
GAO	General Accounting Office (USA)
GATT	General Agreement on Tariffs and Trades
GB	Great Britain
GEC	General Electric Company PLC
GOV	Government
GPS	Global Positioning System
HSCT	High Speed Civil Transport
HSR	High Speed Research Program
IBM	International Business Machines Corporation
IMP	Industrial Modernization and Incentive Program

INTA	Instituto Nacional de Técnica Aeroespacial "Esteban Terradas"
IPR	Intellectual Property Rights
IT	Italy
LAI	Lean Aerospace Initiative
LaRC	Langley Research Centre
LCA	Large Commercial Aircraft
ManTech	Manufacturing Technology Program
MBB	Messerschmitt Bölkow-Blohm
MDC	McDonnell Douglas Corporation
MES	Minimum Efficient Scale
MIT	Massachusetts Institute of Technology
MMC	Metal Matrix Composites
NACA	National Advisory Committee for Aeronautics
NAS	Numerical Aerodynamical Simulation
NASA	National Aeronautical and Space Administration
NIVR	Nederlands Instituut voor Vliegtuigonwikkeling en Ruimtevaart
NL	Netherlands
NLR	Nationaal Lucht-en Ruimtevaartlaboratorium
NRC	National Research Council
NSF	National Science Foundation
NSTC	National Science and Technology Council
ONERA	Office National d'Etudes et de Recherches Aérospatiales
OSTP	Office of Science and Technology Policy (USA)
PMCs	Polymer Matrix Composites
R&D	Research and Development
RD&P	Research Development and Production
RDT&E	Research, Development, Testing and Evaluation
RLV	Re-usable Launch Vehicle
R&PM	Research and Program Management
R&T Base	Research and Technology Base Program
R&T	Research and Technology
SBIR	Small Business Innovative Research Program
SDI	Strategic Defense Initiative
SE	Sweden
SNSB	Swedish National Space Board
SRON	Stichting Ruimteonderzoek Nederland
SSC	Swedish Space Cooperative

SST	Supersonic Transport
STTR	Small business Technology Transfer Research Program
TRP	Technology Reinvestment Policy
UK	United Kingdom
U.S.	United States
USA	United States of America
USAF	United States Air Force
USTR	United States Trade Representative
VAT	Value Added Tax
VITAL	Vehicle Management Integration Technology for Affordable Lifecycle Cost
VMS	Vehicle Management System
WTO	World Trade Organization

1 Introduction: Explanation and Scope of the Study

Purpose of the Study

This analysis of the U.S. large commercial aircraft (LCA) manufacturing sector aims to provide greater detail and accuracy than previous studies on this vexed question (see chapter two). The information provided here is critical to the LCA trade debate, as since the beginning of the 1990s, the U.S. government has felt able to increase federal support for the large commercial aircraft sector, seemingly without fear of invoking a legal challenge under the ensemble of GATT regulations, which apply to LCA trade. In short the U.S. authorities seem confident that billions of dollars of support for the American aerospace industry, including the LCA manufacturing sector, does not constitute subsidy. The analysis here will contest this assumption and will show that subsidy does indeed exist. This study is being written to widen and deepen the European understanding of the U.S. LCA manufacturing industry and to seek to trace with greater accuracy the flows of government financial support which go to U.S. LCA manufacturing companies, notably the Boeing Corporation, which absorbed its one U.S. rival, McDonnell Douglas, in 1997.

The trade friction between the EU and U.S. in the large commercial aircraft sector concerns rival allegations regarding public subsidy of the LCA industry. For the European side a continuing problem has been a refusal by the Department of Commerce, the USTR and U.S. industry executives to acknowledge any government support of U.S. LCA programs. However, this study will show that the U.S. LCA sector is the recipient of substantial federal subsidy and will further detail its form and quantity. In this chapter we will introduce and outline a number of arguments that are fundamental to proving the existence of subsidy. Our focus here concerns the following:

➢ Definitions and explanation of the R&D process
➢ The nature and definition of subsidy

> ➢ The purpose of NASA and DoD R&T programs
> ➢ The existence of defence/civil synergy.

R&D: A Note on Terminology and Definitions

This study seeks to show how the U.S. LCA manufacturing sector benefits from the research and development funded by the American federal government. In layman's terms R&D is the generic concept that applies to the overall process we have investigated in this study of U.S. subsidy. However, within the aerospace industry, government offices and research establishments, the R&D process is often referred to as Research & Technology (R&T). In Europe, development is often designated as a separate activity, which commences when components, systems and platforms are developed towards actual production. The problem, though, is that R&T is actually an early part of the product development process, which is why the generic term R&D is not invalid. To complicate matters further, the U.S. DoD calls its R&D process, Research, Development, Testing and Evaluation, (RDT&E).

To give guidance to the reader we outline below the definitions of the R&D process we have utilized. In this study we have adopted the terminology of the *NASA/DoD Aerospace Knowledge Diffusion Study* conducted by Thomas Pinelli et.al, which construes R&D as divided into three distinct phases:

R&T development
Technology demonstration
Systems development.

These three phases represent an interrelated set of activities in the R&D process where technology is conceptualized, demonstrated, produced and then fed into systems development. However, a key point about R&T and technology demonstration is that definitions refer to the intended outcome, not the physical mechanisms involved in R&D, (Pinelli et. al. 1997, p. 111). R&T consists of the first two phases listed above. Systems development consists of activities aimed at producing aircraft systems, or systems intended for use on a particular platform. In this report NASA activities will be referred to as R&T, as technically they conclude before systems development commences. However, a key point to bear in mind is that NASA programs emphasize technology demonstration. This is a vital

2

support to U.S. LCA because demonstration involves testing configurations that are very close to the final intended market application. In other words demonstration bridges the gap between R&T and development and is therefore of direct benefit to private companies. Demonstrators often lead to significant savings at the development stage, thus reducing financial risk to the companies. As the NASA/DoD *Aerospace Knowledge Diffusion Study* notes, the U.S. LCA sector is closely involved in the large scale testing of technology with NASA and the DoD, (Pinelli et. al., 1997, p.112).

U.S. Department of Defense Definitions of R&D

NASA R&T corresponds to DoD budget classifications 6.1 (research), 6.2 (exploratory development) and 6.3A (advanced engineering). However, DoD programs extend to 6.3B and beyond, which is where development begins. For a full listing of DoD R&D categories the reader is referred to Appendix F. In this study, we adopt the following rule. DoD R&D will be termed RDT&E, NASA R&D will be designated Research and Technology (R&T), as it is pre-developmental, (according to European usage). However, in keeping with normal usage we will refer to the generic process as R&D. We can only apologize to the reader for the fact that sources we quote may use other terminology, such as Research Development and Production (RD&P), (Pinelli at.al 1997).

NASA and DoD Mechanisms of Federal Support to U.S. LCA

In the United States the two major mechanisms by which the federal government provides financial support to the LCA sector are via NASA R&T and DoD RDT&E programs. The DoD also provides additional support through a number of other mechanisms. The following activities and processes constitute the basis of NASA and DoD supports:

NASA:

➢ Facilities, materials and personnel funded by the U.S. government whose function is to ensure the continued world leadership of U.S. civil aeronautics

➢ Generic research undertaken in the Research and Technology Base Program to provide long-range technology transfer into U.S. LCA programs

➢ Research in the Aeronautics Focused Programs to provide technology transfer into U.S. LCA programs which are already in a multi-definition concept phase

➢ Technology demonstrators to bridge the gap between conceptual analysis and real components, systems or platforms

➢ Transfer of skills and technology to U.S. LCA sector employees working on NASA programs.

DoD:

➢ Procurement of military aircraft where airframe technology, avionics, flight management systems and propulsion technology migrate to civil aircraft

➢ R&D contracts for military aircraft systems and platforms that have spin-off potential

➢ Independent Research and Development funds as an overhead on procurement contracts

➢ Military procurement as a buffer against downturns in the commercial market

➢ Dual-Use initiatives to promote the development of commercial technologies that can be leveraged into defence systems

➢ Manufacturing Technology (ManTech) programs to improve production efficiency in prime aerospace companies and the supply chain

➢ Training of engineers on defence projects who transfer to civil programs.

The list of support mechanisms listed above is not exhaustive, other supports come from tax concessions, financial support for exports and executive level co-ordination of sales campaigns. In addition the departments of Commerce and Energy undertake technology initiatives relevant to the U.S. LCA sector.

The Function of NASA R&T

It is now widely recognized that knowledge is a core element in corporate competitiveness. As Golich and Pinelli comment, 'Knowledge is the foundation upon which researchers build as they innovate. Knowledge lies at the core of a state or firm's ability to survive in a competitive world', (Golich and Pinelli, 1998, p.1). The wider understanding in society and the business world that knowledge is now the key corporate asset has come about rather slowly. However, in the last few years the media and serious management journals have come to recognize the strategic role of what is now called "knowledge management", (Schendel, 1996, p. 166). As a number of studies have noted, knowledge is often situated outside the firm and must be diffused across institutional boundaries, (Lynskey, 1999, p. 317).

Knowledge management necessitates knowledge creation and knowledge dissemination. The role that NASA plays in creating and disseminating aeronautical knowledge warrants explanation. A core requirement for commercial competitiveness in aeronautics is now knowledge. In order to remain competitive in LCA manufacturing, companies require that their core competence in technical knowledge be constantly upgraded. As Phil Condit, CEO of Boeing, succinctly noted in 1997 '...you fail to innovate you lose', (quoted Golich and Pinelli, 1998, p.1). This requires an intensive effort in research, technology and development, as well as the ability to move knowledge between institutions. The need for state of the art technology in civil aeronautics comes from the commercial pressure on the LCA business to accelerate the aircraft development cycle and to bring advanced technology to market more rapidly. Further, new products must have reduced direct operating costs (DOCs). As we will show this was achieved by Boeing on its B-777 program with federal assistance. The strategic role of public agencies in this process, such as NASA, is to fill the funding gap between what is required in R&D and what is actually funded by private companies. As the 1997 *NASA/DoD Knowledge Diffusion Study* notes:

> In the case of LCA, where RD&P costs are very high, technological innovation is virtually impossible for individual firms to appropriate, and return-on-investment cycle - assuming they are successful – can take as long as 10-15 years, creating demand may not be enough to impel firms to innovate. National governments with active LCA producers... all invest

> public money in an effort to underwrite the risk because LCA manufacturing is a strategic industry sector, (Pinelli, et.al, 1997, p. 65).

Federal funding of aeronautics R&T can thus be construed as a subsidy to companies to assist them in remaining at the forefront of technological knowledge. At NASA, the corporate relevance of this knowledge is underscored by the fact that the bulk of the aeronautics $1 billion per year expenditure on R&D contracts (75%) is awarded to private companies, (ASRC analysis). More pertinently, 65% of these contracts go to aerospace companies. Thus, as well as the research and technology capability that results from these contracts which can be exploited in current and future applications, it must be realized that this process actually trains and reskills the workforce of the companies involved in U.S. LCA manufacturing.

The Nature of Aeronautics Knowledge

In order to be competitive as an LCA producer, companies require different kinds of knowledge to be embedded in their core competence. Such knowledge is part of what economists call the "intrinsic" competitiveness of a company. Intrinsic competitiveness is embedded in the technological and organizational capability of firms. Knowledge is also now fundamental to the central mission of LCA producers, because it is the key resource in reducing technological and financial risk. According to Pinelli at.al. public funding of R&T enhances all the four categories of knowledge listed below, (1997, p. 63).

➢ Product embodied knowledge
➢ Process embodied knowledge
➢ Systems integration knowledge
➢ Management knowledge.

Product knowledge is embodied in the components, systems, materials and structures within an aircraft. Thus new avionics systems and components, or new composite materials embody product knowledge.

Process knowledge resides in the know-how and some of the manufacturing processes of LCA. It is difficult to quantify, as it is often tacit and not amenable to formal delineation. In engineering circles, process knowledge is closely associated with the notion of "engineering judgement".

Systems integration knowledge refers to the overall capability to integrate successfully the exotic materials, systems and components that make up a modern civil jetliner, a product now more structurally complex than a nuclear reactor. In LCA, co-ordination and execution of final assembly embodies systems integration knowledge.

Management knowledge refers to the capability of management to organize projects, lead teams and orchestrate effective co-ordination and communication. In the LCA sector a key use of management knowledge is controlling the overall value chain implicit in the supply chain of vendors and suppliers.

In the U.S. LCA industry, companies have benefited in the growth of all these forms of knowledge, from public funding. However, the key benefits to the companies have been found in product and process knowledge. Systems integration knowledge and management knowledge are more properly located within the domain of the company, although cases can be found where public funding of research on the overall manufacturing process is in evidence. For example this has been a long-term goal of the DoD's ManTech program, (National Research Council, 1999, p. 17). A more recent example is the USAF-sponsored MIT Lean Aerospace Initiative (LAI), which is a consortium of academia, industry and government that is seeking to revolutionize systems integration capability and management knowledge in U.S. aerospace companies, (http:/lean.mit.edu/public/welcome.html, 21/8/98).

Product knowledge and process knowledge are critical resources for U.S. LCA manufacturers, and it is here that government funded NASA programs confer a direct benefit to the U.S. manufacturers through the upgrading of knowledge competencies. As we will show in chapter three the NASA aeronautics R&T Base Program explores basic research which would be ignored by private companies because of cost factors. From the R&T Base Program a long stream of technical innovations have been developed which have fed into specific American LCA programs, (NASA Office of Aeronautics, FY 1996 Budget Report).

In the Aeronautics Focused Programs, such as the Advanced Subsonic Transport (AST) and the High Speed Research Program (HSR), R&D is closely tied to more immediate commercial goals, such as cheaper means of producing composite materials in order to reduce the overall weight of new airframes and hence ensure lower operating costs. As a U.S. Council on Competitiveness report indicates, the focused programs are closely linked to commercial pressures in the LCA sector:

> NASA undertook these efforts [HSR and AST] largely in response to fears that the U.S. aircraft industry was falling behind Airbus in its technological capabilities, as well as to help industry address the gap that had emerged in its commercialization of new technologies...In addition, NASA is trying to make its own research efforts more responsive to the commercial manufacturing cycle by timing the development of new technologies to coincide with the onset of commercial programs, (Council on Competitiveness Report, April, 1996, p. 33).

Defence/Civil Synergy

In the United States there has been a growing assumption in recent years that defence/civil synergy in aeronautics is in decline. This assumption is very convenient because it implies that federal funding of defence provides no benefits to the civil side of the industry. However, this assumption is wrong. The U.S. Department of Defense supports the U.S. LCA industry via the spin-off potential of defence R&D and the synergy between defence and commercial components, systems and platforms. As Roland notes, '...But as a rule fundamental advances in one realm have been applicable to the other...There is simply no clear dividing line between civilian and military aviation', (Roland, 1978, p. 366).

A key assumption of this study on subsidy of U.S. LCA is that aeronautical R&T is generic. Only different functionalities create the division between civil and military aircraft. Nevertheless, these functionalities can be highly diverse. But what is often overlooked is that differing engineering objectives can be achieved with generically related technologies. The avionics of a fighter are radically different from those of a large commercial aircraft. But today both will share some generic technologies related to computer controlled flight management systems, with an avionics architecture consisting of highly integrated modular components. Thus, although we readily concede that the functionalities of a modern fighter and LCA are divergent, even here there are inter-relations and synergies in avionics. For example, the Boeing 777 has a state of the art avionics suite based around the Honeywell Aircraft Information Management System (AIMS). However, it was the U.S. Air Force that pioneered the first generation of digital avionics systems for the F15 and F16 fighters. In the 1980s a joint services initiative pushed R&D towards the modular avionics concept with the Pave-Pillar and Pave-Pace programs, which were undoubtedly the pre-cursors to the B-777 system.

A second area where defence/civil synergies exist is in materials R&D. The LCA industry today is facing acute pressure from customers for significant operating cost reductions, particularly in fuel savings and reduced maintenance costs. For fuel savings to be achieved it is necessary to reduce weight by replacing traditional metal alloys with composite materials, particularly polymer matrix composites, (PMCs). Already 10-15% of the structural mass of a large civil jet can consist of PMCs, but a number of manufacturing challenges have held back greater use of composites (SRI). However, for the civil sector, the affordability of composites has been made possible by military-funded programs, which have sought to reduce the manufacturing costs of these exotic materials. As we will show in chapter four Boeing's work as a subcontractor on the Northrop Grumman B-2 bomber program enabled the company to develop expertise in the use of composites which was fed across to the B-777 program. According to *Aviation Week* journalist William Scott the Boeing Corporation received $100mn from the DoD's ManTech program for the development of innovative processes in the manufacture of the bomber, some of which migrated across to the B-777, (Aviation Week and Space Technology, 5/12/88).

As well as polymer matrix composites, new kinds of metal alloy are finding their way into large civil aircraft production, including metal matrix composites, (MMCs). Again, public funding has been decisive in making key breakthroughs in titanium and superalloy manufacturing. According to the U.S. National Research Council the ManTech Program provided industry with isothermal forging and net-shape manufacturing of superalloys, (National Research Council, 1999, p. 18). On the B-777 new Titanium Beta alloys developed during the NASP X30 program have a number of applications on the engine nacelles, (*Journal of Metals,* May, 1992). Similarly, Titanium 10-2-3 has been used on both the B-777 and C-17 landing gear, (*Mechanical Engineering,* July 1993). Again, the key point regarding these alloys is that they are both costly and difficult to manufacture, hence DoD funded programs to improve manufacturing efficiency, such as ManTech, are making available technology for U.S. LCA that the private sector cannot afford to fund independently.

We have cited these examples of synergy here in order to counter the growing assumption in U.S. policy-making circles that defence/civil synergy is in decline. At the level of the whole airframe, fundamental

9

synergies continue to exist between heavy-lift transports, tankers and civil jets. But even with modern fighter aircraft, potential spin-off exists from materials and avionics to civil applications. The fact of defence/civil synergy is underscored by the number of dual-use initiatives currently being pursued in the U.S. As we show in chapter four these aim to leverage civil technologies into the defence sector and to reduce the manufacturing costs of defence related civil technologies. If synergy were now in decline then these initiatives would not make sense. Table 1.1 illustrates the possible areas of synergy and spin-off.

Table 1.1   Defence/Civil Spin-off Potential

	Defence/Civil Synergy
Aerodynamics	Computational Fluid Dynamics, Aerodynamical Simulation and Design Tools
Airframe	Tanker/Heavy-Lift/Freighter
Avionics	Integrated Modular Avionics, Fly-by-Wire, Fly-by-Light, Flat Panel Displays, Mission Management etc
Design	Computer Aided Design, Computer Aided Manufacturing Tools
Manufacturing	Lean Manufacturing, Laser Machining, Computerized Numerical Control, Composite Tape-Relay Machining, Isothermal Forging
Materials	Composites, Aluminium & Titanium Alloys, Metal Matrix Composites

Dual-use and synergy is also very much targeted at the supply chain. With increasing outsourcing of aircraft development and production, spin-off potential exists from a number of suppliers and vendors, whose components and systems, developed through government funded programs, may find their way onto U.S. LCA. Similarly with more than 50% of the value of the whole aircraft now outsourced by Boeing the flow of subsidy must be seen to pass through vendors and suppliers, (Pinelli et. al, 1997). As table 1.1 above indicates a number of areas continue to exist where components and systems developed for military applications can migrate across to the civil side. In the case of tankers and transport aircraft synergy even exists at the whole aircraft level. As we write Boeing is preparing to launch the MD-17 freighter derivative of the C-17 military transport it inherited from McDonnell Douglas. The proposition that Defence/Civil synergy is in decline is thus open to serious question. The truth is that the links between civil and military programs are undoubtedly becoming more complex and difficult to track. However, if synergies are now of a lesser magnitude one must ask what was the rationale for Boeing's 1997 acquisition of McDonnell Douglas? As we will indicate below the recent policy of the Clinton Administration has been to bring the civilian and defence sectors of aerospace closer together. Again one questions how this can make sense if the two sectors are now so divergent. To repeat a fundamental point the assumption of declining defence/commercial synergy has been significantly overstated.

Conclusion

The American federal government subsidizes U.S. LCA manufacturing through a variety of levers, chiefly operated by NASA and the DoD. The purpose of the various forms of subsidy is to enable the U.S. to remain competitive in the global LCA industry. NASA's R&T programs represent a subsidy to the U.S. LCA sector because they generate knowledge and technology transfer to the American LCA industry. We contend that the public funding of these R&T activities exists because the private sector is unable or unwilling to provide funds for this kind of research. A key resource in corporate competitiveness is now knowledge and NASA programs explicitly seek to enhance the knowledge base of U.S. companies in order to promote their global competitiveness. In addition, NASA programs reduce the interrelated technological and financial risk associated with new LCA programs, (for a full discussion of NASA programs see chapter four).

The enrichment of corporate knowledge is achieved through technology transfer from NASA and the DoD to the U.S. LCA sector. In the case of NASA, R&T is transferred to the U.S. LCA sector as a supplement to in-house R&T. The DoD situation is more complex. In this case the advantages going to the U.S. LCA sector derive from the fact that military and civil manufacturing are closely linked and often simultaneous. The aim of DoD dual-use programs is to improve the technological capability of the U.S defence/industrial base. But public funding of defence technology programs or manufacturing infrastructure improvements inevitably benefit firms which are both civil and defence producers.

In subsequent chapters, by using the approaches outlined in our methodology, we show clearly the quality and quantity of the subsidy to the U.S. LCA industry that derives from NASA and DoD programs. This analysis is provided in chapters four and five. The detailed account of the underlying methodology can be found in Appendix A.

2 The Large Commercial Aircraft Trade Issue in Historical and Theoretical Perspective

A Note on Trade Theory

According to the now dominant neo-classical (liberal) school of economics the ideal structure for world trade is one based on open and free markets. Free trade is based on the efficiency maximizing principle of comparative advantage. The logic of comparative advantage is disarmingly simple: a country should produce and export goods which it can manufacture more efficiently than its trading rivals; alternatively it should import goods produced with greater efficiency by other countries. From a liberal point of view the aim' is to secure an international division of labour based on economic efficiency. Because of factor endowments in capital, labour and land particular countries will have a "natural" advantage in some economic sectors, which they should exploit.

The theory of comparative advantage, developed by liberals such as Ricardo, Smith and J.S. Mill, undoubtedly captures an abstract principle of economic efficiency. However, when the theory collides with the real world significant problems emerge with its application. National economies exist within the boundaries of nation states which have their own economic agendas. In modern industrial societies governments have been acutely conscious of a strategic hierarchy of industries which are ranked in terms of factors, such as degree of value-added by labour, prestige and contribution to national security. In particular states have been anxious not to be wholly dependent on trading partners for industrial technology they regard as critical to their science/technology base and national security. Thus in the real world nations have pursued strategies to catch-up and overetake other countries in critical industrial technologies. So called welfare economists have attacked this practice as leading to distortions and inefficienies, but this is whistling in the wind. Seen from a mercantilist political economy perspective industrial policy strategies to bolster and support certain industrial sectors make eminent sense. Empirically, world trade follows much more a drive to secure competitive advantage, rather than comparative advantage. The point is that through the use of certain policy instruments the framework of relative advantage in industrial trade can be altered through conscious government and corporate policy. Moreover, governments have a political responsibility to their citizens working in industrial sectors which are subject to foreign competition. They are required to lay down a framwork for the economy which strengthens the overall competitive position of the nation. In addition internal restructuring of the national labour market, which might be indicated by comparative advantage, will often be impossible for legitimate politcal reasons.

The reality of strategic trade, as opposed to free trade, is now widely recognised, but still hotly contested. After the successful catch-up strategy of Japan in consumer electronics and autos, inspired by the activities of the Ministry for International Trade and Industry (MITI), economists began to see the advantages which could accrue fron effective industrial and trade policy. Balaam and Veseth put it neatly; 'The point is that in a more competitive environment, comparative advantages are no loger fixed but can be manufacture by states and firms', (1996, p.23). Thus rather than accept an accidental distribution of productive advantage states have sought to improve their capability in high-technology/high value-added industries, such as aerospace. As the perception of the role of strategic trade grew clearer the world's leading industrial nations became entangled in a series of bilateral and multilateral negotiations and disputes over how to secure "fair play" in international trade; "fair trade" replaced free trade.

In January 1995 these negotiations culminated in the creation of the World Trade Organisation (WTO), which now has global judicial authority over its members' trade. Sructurally and ideologically these negotiations have been shaped by the U.S. pressing Japan and later the EU over trade issues. In the European case aeronautics has been a major area of concern to the U.S., as Airbus Industrie has eaten into the U.S.'s world market share, in addition all but one of the U.S. prime companies has left the commercial aircraft arena. However, as the case of aeronautics illustrates, the issue of a level playing-field for trade is extremely complex. What needs to be understood is that free-market ideology is itself a resource in the propaganda battle over trade issues. The United States, which has initiated much of the process of negotiation and organised most of the relevant international machinery, perceives itself to be the bastion of free-trade. However, in the view provided here, the U.S. has been relatively unconscious of the role of its own government in supporting strategic industries. In our view the liberal characterization of the U.S.

economy in the post-war period is exaggerated and overblown. Some sectors, such as consumer electronics, have felt the chill wind of global competition, others have not. Protection has been apparent in agriculture at one extreme and aerospace at the other. Judith Goldstein has outlined 3 models of trade relations based on an analysis of ideological and institutional factors which determine the prospects for political support. She conceives fair trade and redistributive trade as additional models to the orthodox free trade mantra, (1988, p. 216). But based on our realist perspective we would add a fourth category of strategic trade in products centrally linked to issues of national security. Post-1945 U.S. trade in defence and aerospace has never simply been about commercial factors. Civil aeronautics exports have been closely linked to security issues and on occasion have arisen on the back of defence sales. In terms of assessing which industries are likely to receive aid and protection Goldstein noted how those prone to unemployment and those considered successful and highly competitive were most likely to obtain high level political assistance. Again aerospace fits this categorization. As we shall argue below the more overt mercantilist policy of the Clinton era is related precisely to the competitive challenge mounted by Europe in civil aeronautics in the last 20 years and the high levels of unemployment experienced in the U.S. industry after the end of the Cold War. According to Jens Van Scherpenberg the neo-mercantilist aspects of this process are becoming more pronounced as the constraints on conflict with allies in Europe are loosened. In his view aerospace exports have been at the centre of an executive steered advocacy policy which is seeking the rents and positive externalities due to the state which bankrolled the West's Cold War security policy. As Van Scherpenberg notes, 'Linking military dominance with an aggressive pursuit of economic interests has since become a core element of the US economic policy agenda', (1997, p. 107).

We outline here the nature of U.S. government supports precisely because the mainstream academic theories and much political commentary uncritically underscores the U.S. position.

Documenting LCA Subsidy

The global market for large civil aircraft and the strategic behaviour of key players within that market have become the focus of considerable political and corporate concern over the last decade. In the last 20 years the share taken by U.S. companies of new orders for large civil aircraft in the global market has declined from roughly 95% to about 60% with the European partnership,

Airbus Industrie, increasing its share to 40% of new orders. More dramatically, in 1999 Airbus shocked its U.S. rival by capturing 55% of the market. Consequently, attention has focused upon the erosion of the dominant position of the U.S. manufacturers in the global market for commercial airliners, as a result of the increased market penetration of the European Airbus consortium (from 2001 a jont venture company). The increasing market success of Airbus Industrie during the 1990's, which coincided with a period of sharp contraction for Boeing, has intensified concerns in the U.S. about the competitive nature of the market and the degree to which overt government support for overseas competitors in the aerospace sector is acceptable. As we have already noted this is in a context where the U.S. tends not to see its own funding as "subsidy".

The intensification of the trade dispute between the U.S. and EU during the early 1990's was been conducted in the public arena through a series of commissioned reports: namely Gellman Research Associates (1990) and Arnold and Porter (1991), emanating from both sides. In addition trade officials on both sides have monitored each other with confidential, unpublished studies. Each of the published reports served a specific objective. Gellman enabled the U.S. to question the business viability of the Airbus Industrie program. The report asserted that Airbus was only viable with substantial subsidy from participating European governments and that such support was inequitable. Thornton comments that the Gellman report, 'clearly staked out a U.S. position that took a very dim view of Airbus and the government money supporting it', (Thornton, 1995, pp.140 - 141).

The 1991 Arnold and Porter report represented a detailed response to the U.S. position from the European Commission. This study attempted to document the scale of U.S. direct and indirect subsidy to its aeronautics industry (particularly that flowing through Department of Defense military contracts and NASA aeronautics industry support) and implied that financial support for the European aerospace industry (principally in the form of repayable launch aid) was merely a necessary response to earlier and significant intervention by the U.S. government to protect its industry. With reference to the arguments proposed by the Arnold and Porter report, Thornton comments that: ' ...most observers of the aerospace industry agree that military programs subsidize civilian ones in a variety of ways. This is especially significant regarding the steady stream of financing provided by military contracts, which help to fund the long and risky development of civil products', (1995, p.145).

The Gellman report and the Arnold and Porter response clearly set the scene in the early 1990s for the continuing transatlantic trade dispute concerning LCA that was to unfold over the rest of the decade. It is the purpose of this section of oiur study to review the findings of both Gellman and Arnold and Porter in order to provide a backdrop to the debate and a starting point for the ensuing analysis.

The Gellman Report

In September 1990, Gellman Research Associates, of Jenkintown Pennsylvania, issued a report entitled: *An Economic and Financial Review of Airbus Industrie,* which had been prepared for the International Trade Administration Division of the U.S. Department of Commerce (DOC). The report focused upon the economics of Airbus Industrie's civil aircraft programs and examined the potential effects of Airbus Industrie's activities on both the market for civil transport aircraft and on competing U.S. firms. The Gellman team made clear from the outset that one of the primary reasons for undertaking the study was to compile and assess data and information on the Airbus Industrie operation and the levels of government support that its program received. Gellman suggested that such a study was necessary because the Airbus partnership did not make available such detailed information on its financial performance, nor on the support it received from the partner's individual member governments. Furthermore, the Gellman team were tasked with clarifying 'the complex web of relations between the participating companies, the governments and the AI consortium', (Gellman, ES1). At the outset, however, Gellman emphasized the existence of 'a degree of uncertainty in the numeric estimates presented', (1990, ES1). This was appropriate as their calculations were necessarily based upon speculative price, quantity and cost estimates for past, current and future Airbus Industrie aircraft sales.

The Gellman team drew a number of important, although contentious, conclusions from their research and, in presenting an essentially negative view of the market role and performance of Airbus Industrie, made necessary a powerful response from the European consortium and effectively acted as the catalyst for the trade friction which has followed. The Gellman report contained four key findings, presented as apparent allegations about the market performance and financial viability of Airbus Industrie.

First, Gellman called into question the commercial viability of Airbus Industrie programs. The allegation was made that such programs, either individually or together, could not demonstrate past, present or future commercial viability. It was claimed that all Airbus Industrie programs revealed negative net present value, when commercial borrowing costs in Europe (averaging around 8.7% at the time) were employed to discount cash flows. The implication drawn by Gellman was that no privately financed firm would have invested in any of the Airbus programs, concluding in turn that the Airbus partnership could only function with significant government support.

Gellman examined the discounted cash flow issue at the level of individual Airbus programs to illustrate the point. For example, Gellman alleged that the original A300B aircraft experienced significant negative cash flows, even when supported by government launch aid. Gellman suggested that these financial losses have been compensated for, in part, by additional government production subsidies and equity injections. Furthermore, Gellman implied that, on their estimates, the 1977 A310 program would also experience significant negative cash flow, excluding any interest charges, amounting to some $12.9 billion in 1990 prices from its introduction until 2008. The 1983 A320/A321 program was expected by Gellman to yield a negative nominal cash flow of $4.9 billion in 1990 prices from introduction up to 2008. However, Gellman indicated that their estimates suggested that the 1987 A330/A340 program would show a positive nominal cash flow of $3.2 billion in 1990 dollars to 2008, although it would still not produce a positive net present value using commercial rates of interest, relevant at the time of the report. The Gellman projections for the then most recent Airbus Industrie programs, the A320/A321 and A330/A340, suggested that, in the strengthening global market for large civil aircraft, there was a much improved chance of financial viability and, as a result, Gellman argues that the case for additional government support would be thereby negated.

Secondly, Gellman focused upon the issue of government subsidy directly and attempted to identify and measure such support for Airbus Industrie ventures emanating from European governments. Up to 1990, Gellman estimated that the governments of France, West Germany and the United Kingdom provided some $8.2bn in support to Airbus Industrie partner companies. In addition, a further support program amounting to some $2.3 billion had been pledged for the A330/A340 program. Furthermore, it was argued that the merger of Daimler-Benz and MBB, the parent company of Deutsche Airbus, provided an additional \$3bn in financial support. Gellman asserted that the governments of the Airbus Industrie partners had provided about 75% of aircraft development funds over the lifetime of the partnership which, had commercial rates been applicable, would have amounted to some \$25.9 billion in 1990. Controversially, the Gellman team concluded that their financial analysis of the partnership in 1990 implied that future profit generation would prove insufficient to ever repay government support in full. Despite being able to charge higher prices in the improving market conditions of the late 1980's the conclusion was that genuine commercial viability could not be attained over the next two decades.

Thirdly, the Gellman report drew attention to what they perceived to be the market impact of the Airbus partnership. Their starting point was the observation that only a limited number of privately financed firms can exist in the market for specific types of transport aircraft. Even in a growing global market, international demand for individual types of aircraft is inevitably limited to a few hundred units each year. Initial launch costs are usually very high and the economics of the industry is such that average unit costs of production fall as output increases and production levels approach the minimum efficient scale (MES). To survive and prosper in the industry, firms have to succeed in selling enough units to take advantage of such economies of scale as they move down the learning curve. This is vital in the aerospace industry as learning elasticity is of the order of .2, thus a doubling of output can lead to a 20% reduction in production costs, (Klepper, 1990, p. 777).

Gellman argued that Airbus Industrie (and its individual partner companies) had been able to overcome prohibitively high entry costs into the large civil aircraft industry as a result of significant levels of past and present government financial support, which also enabled the partnership to withstand more easily market pressures than could privately-financed companies. Gellman made the point that: 'so long as AI partner companies continue to receive subsidies from their governments, AI can continue to compete effectively without the necessity to make its programs financially viable', (Gellman, 1990, ES3.)

Gellman noted that the commitment of government support to the European aircraft industry had much to do with the significant external benefits associated with its activities, such as the expansion of high-technology employment opportunities and the stimulation of other advanced technology industries. In addition Airbus was associated with the policy of trans-European co-operation that the partnership represented at a time when Europe was seeking closer industrial integration.

Finally, the Gellman study turned to the impact of the Airbus partnership on the industry in the USA. Rather than recognizing that Airbus Industrie had seized the competitive edge in the global market for large civil aircraft through highly efficient production and the application of leading-edge technologies, Gellman blamed the loss of market share by the U.S. producers on European government support for Airbus. Stating categorically that Airbus Industrie 'cannot exist' without such support, Gellman anticipated continued decline in the U.S. share of the LCA market and associated profits as long as Airbus would be able to sell its aircraft below cost, protected by government subsidies, (1990, ES3).

Gellman, accepting the questionable assumption that U.S. aircraft manufacturers operate without direct or indirect government support, suggested that declining profits for U.S. manufacturers would constrain their investment from internally-generated funds and potentially discourage the development and introduction of new U.S. aircraft programs. Alternatively, U.S. manufacturers would have to seek alternative sources of funding through foreign investment, potentially putting U.S. technological edge at risk through technology transfer.

Overall, the Gellman report made very clear that the U.S. viewed the emergence and recent remarkable market performance of Airbus Industrie as a major threat to its former dominance of the LCA industry and attributed blame for loss of market share to the subsidy strategy they perceived to be implemented by European governments.

Regarding the validity of the Gellman report's assumptions a number of key criticisms can be made.

1.   The market forecast used to predict future Airbus sales and returns have turned out to be wholly wrong. The A320, which entered service in 1988, proved to be the fastest selling commercial aircraft in history. Repayments of launch aid for the A320 family are already showing a healthy return for the investments of the respective governments.

2.   Gellman's discounted cashflow model was used to inflate the values of Airbus subsidy by applying 'commercial rates of interest' to government loans. However, the empirical values for interest rates applied to industrial investment vary enormously depending on the risk assessment, the base rates of the country

involved and the nature of the lending institution. In other words it is not simply a mechanical task to ascertain historically valid commercial rates of interest.

3.   The Gellman report really implied that Airbus ought not to exist because the free market would not have brought forth the requisite investment in civil aircraft production. This is spurious and shows a wholesale disregard for the economics of the LCA industry. The commercial jetliner industry in the USA grew out of military funded programs. Quite simply the "start up" costs of U.S. LCA were borne by the DoD. Throughout its history, aeronautics has occupied a crucial nexus connecting the economic and military dimensions of national security, and thereby combines the commercial and political in a unique fashion; it is the quintessential strategic industry. The notion of a laissez-faire aeronautics industry is a fairy tale.

The Arnold and Porter Report

Following Gellman's direct and uncompromising attack on the economic and commercial viability of the European aircraft industry, it was imperative that Europe produced a cogently argued response. The Arnold and Porter report, *U.S. Government Support of the U.S. Commercial Aircraft Industry*, was prepared by the Washington D.C.-based consultancy for the European Commission and issued in November 1991. It set out to undermine the conclusions of the Gellman report by offering a definitive picture of the significant degree of financial support historically and currently offered to U.S. aircraft manufacturers by the U.S. government and thereby challenging Gellman's "privately-financed only" view of U.S. industry. Arnold and Porter thus amplified the criticisms we have already outlined above.

The Arnold and Porter study was equally as direct as Gellman in the assertions it made about the U.S. LCA industry and the support it receives from government, albeit in this case indirectly. Arnold and Porter's research identified 'massive systematic support to the U.S. commercial aircraft industry pursuant to a long-standing U.S. policy of striving to maintain U.S. superiority in all areas of aeronautics technology', (1991, p. 1). In common with the Gellman report, Arnold and Porter draw attention to the lack of transparency in the data available upon which to base their research.

21

Despite the data problems identified, Arnold and Porter estimated that U.S. government support to its own commercial aircraft industry in the period 1976 to 1991 was in the range of $18 billion to $22.05 billion. Using 1991, rather than historic prices to estimate the value of benefits accruing to the U.S. aircraft industry from Department of Defense and NASA contracts, the estimated range of benefits amounted to between $33.48 billion and $41.49 billion.

Arnold and Porter identified three principal ways in which the U.S. government supports the U.S. commercial aircraft industry. First, substantial support for the aircraft industry in the U.S. is received from U.S. Department of Defense R&D budgets. Such was the over-riding strategic importance of the aeronautics industry to the U.S. in the post-World War II period that vast financial resources were dedicated to military aeronautics R&D. Since the key companies in the U.S. commercial aircraft industry were engaged, directly or indirectly, in military aeronautics development and production and that military and commercial aeronautics technology often overlap, it was inevitable that commercial aircraft development and production would derive very substantial crossover commercial benefits and synergy from their participation in military R&D.

Arnold and Porter cited what they termed "quantum leaps" in U.S. commercial aeronautics technology - the Boeing 707, the wide-body jets and the development of a supersonic civil transport plane - as examples of programs where substantial U.S. government involvement was apparent in the period prior to each breakthrough, (1991, p. 2). They estimated that, since 1976, the U.S. Department of Defense had spent some $50 billion on aeronautics R&D, with at least $6.34 billion reaching the two leading U.S. aircraft producers of large commercial aircraft, Boeing and McDonnell Douglas, to finance aircraft-related R&D. Taking account of the proportion of these funds estimated to have been of benefit to the U.S. LCA sector, Arnold and Porter suggested that it might have derived benefit amounting to between $5.9 billion and $9.7 billion of Department of Defense expenditure. In terms of 1991 dollar value (and accounting for opportunity costs and compound interest), this translates into commercial benefits measured in current prices of between $12.42 billion and $20.18 billion.

Arnold and Porter conceded the point that the Department of Defense did attempt to recapture a proportion of the private commercial benefits accruing to players in the U.S. LCA industry from their involvement in military aeronautics R&D. However, they indicated that in the years between 1976 and 1990, less than $200 million had been recouped, representing a minuscule amount of the total funding committed. It should also be pointed out that DoD recoupment is no longer required from U.S. firms.

The authors also emphasized that, as well as direct Defense Department R&D grants that flow to private U.S. aircraft companies, the U.S. also directly reimbursed these companies for in-house R&D projects that may have military application through the Independent Research and Development Program. Such is the current value of dual-use technologies in the aerospace industry that the commercial utility of such in-house, self-chosen research and development activity is even higher than in government-initiated R&D. In the decade and a half prior to the 1991 report, Arnold and Porter estimated that U.S. aerospace companies had received about $5 billion of such reimbursements from the U.S. government, worth some $1 billion to $1.25 billion of likely benefit to the U.S. commercial aircraft industry.

The second institutional locus of support identified by Arnold and Porter drew attention to U.S. government subsidy for the U.S. commercial aircraft industry, which emanated from NASA budgets. As one of its primary mission objectives, NASA is tasked with the promotion of U.S. technological superiority in the aeronautics sector and therefore provides substantial funding for civil, as well as military aeronautics R&D. Arnold and Porter estimated that, between 1976 and 1991, NASA has committed about $8.9 billion to civil and military aeronautics R&D in the U.S. Much of this expenditure funded large-scale R&D projects of significant value to U.S. civil aircraft manufacturers, for example the Aircraft Energy Efficient Program and the development of the supercritical wing. In addition, funding was provided by NASA for a whole range of smaller projects that focused on the encouragement of specific technological developments in aeronautics.

Taking NASA's role in supporting technological progress in U.S. commercial aeronautics and the inter-linking of its civil and military R&D objectives together, Arnold and Porter estimated that some 90% of NASA R&D expenditure over the period since 1976, amounting to $8 billion, represented a benefit to U.S. commercial aeronautics. At current prices, they suggested that the value to the U.S. commercial aircraft industry of this support would have been almost $17 billion.

As well as assessing the contribution of NASA and the DoD to the U.S. LCA sector Arnold and Porter emphasized the specific ways in which the U.S. tax system provided distinct benefits to the U.S. civil aircraft

23

industry. As they noted: 'the "completed contract method" for determining when contract income is subject to tax has allowed U.S. aircraft manufacturers to reduce taxes by deferring substantial amounts of income. Use of domestic international sales corporations (DISCS) and foreign sales corporations (FSCS) also has permitted substantial deferrals', (1991, p. 5.) The authors estimated that, over the fifteen years to 1991, these important tax deferrals and exemptions benefited Boeing by approximately $1.7 billion and McDonnell Douglas by some $1.4 billion.

Overall, Arnold and Porter suggested that, far from the U.S. civil aircraft industry being a privately funded operation with no government support, at least three identifiable and quantifiable major areas of support existed. The combined value of these government-funding mechanisms had, they asserted, provided the U.S. industry with an estimated commercial benefit of between $18 billion and $22.5 billion between 1976 and 1991. In current 1991 prices, this support would have been valued at between $33.48 billion and $41.49 billion. On top of this massive government support for the domestic aircraft industry, the U.S. government also provided various additional forms of aid including U.S. aircraft manufacturers' use of government test facilities at reduced cost and occasional special purchases of aircraft by government (such as the KC10s purchased by the U.S. government from McDonnell Douglas in 1982).

Arnold and Porter concluded with the observation that, while exact measurement of U.S. government support for its civil aircraft industry is impossible given the lack of transparency in available data, there was little doubt that such support had played a key role in securing and preserving the most important technological advances made by the U.S. civil aircraft industry, helping to secure the position of the U.S. commercial aircraft industry in increasingly competitive global markets.

In summary the Arnold and Porter study served to correct a number of biases and prejudices that previously informed the U.S. view of EU/USA trade in LCA. Chiefly Arnold and Porter demolished the myth that the U.S. LCA industry was an entirely free enterprise operation divorced from the benefits of state support.

The July 1992 Agreement

Despite the intense political rhetoric of the EU/U.S aerospace trade dispute in the early 1990s and the detailed data comparisons contained in the commissioned reports which underpinned it, both sides initially seemed to be prepared to compromise during negotiations in 1992, in order to avoid the dispute spilling over into a major trade conflict in which both sides, potentially, stood to lose. Accordingly, the EU and the U.S. signed the *EU/U.S. Bilateral Agreement on Trade in Large Civil Aircraft* in July 1992. On the European side, concessions ultimately reduced the permitted level of direct governmental support on aircraft programs from 75% down to 33 % of program costs, subject to full repayment on a royalty basis. On the U.S. side, it was conceded that indirect supports were significant in assisting the aerospace industry and, in similar fashion to direct subsidies, ought to be the subject of regulation. According to the 1992 Bilateral the annual value of such support should not exceed 3% of a nation's commercial aircraft industry's annual turnover, or 4% of the annual turnover of any single firm in that nation's aircraft manufacturing industry.

However, as a number of commentators have noted, monitoring U.S, compliance with the 1992 agreement is problematic for the EU, (Lawrence, 1998). This is because there are numerous channels through which the U.S. provides financial support to the LCA sector. In consequence it is difficult to isolate one authoritative source of data on U.S. funding. In addition much of the support is conveyed through R&D contracts where the content of any technology transfer is considered to be proprietary, or in the case of the DoD, top secret. The U.S. authorities have sought to define direct support as identifiable only when demonstrably linked to a specific LCA program. This inflenced the agreement which provides the following definitions of supports:

Indirect support:

Financial support provided by a government or by any public body within the territory of a party, which is provided for aeronautical applications, including research and development, demonstration projects and development of military aircraft, which provide an identifiable benefit to the production of one or more specific large civil aircraft programs, (Annex II, Article 5, 1992).

Direct support:

Any financial support provided by a government or by any public body within the territory of a party which is provided:

1. For specific large civil aircraft programs or derivatives; or
2. To specific companies to the extent that large civil aircraft programs or derivatives directly benefit, (Annex II, Article 6, 1992).

The principal purpose of this uneasy compromise, referred to as the "July Agreement" was to: 'defuse what had become an explosive issue by removing it from the centre of political controversy, thus preventing further complications in already volatile transatlantic trade relations', (Thornton, 1995, p. 146). While the 1992 July Agreement initially served its purpose, the dispute has continued to simmer, both at the political level and between the principal aircraft manufacturers involved. The continued market penetration of Airbus, together with the recent and continuing extensive restructuring of the U.S. aerospace/defence industries poses new threats on both sides and has placed the 1992 Agreement under considerable strain. However, throughout the tenure of the Clinton Administration, it has been chiefly the U.S. side which has sought to re-ignite the issue, with allegations that the EU is in breach of its obligations. In December 2000 the heat in the issue was turned up again when Airbus initiated the industrial launch of the A380 super jumbo, which will end the U.S. monopoly in the 400 plus seat sector of the LCA Market.

The GATT Uruguay Round and the ASCM Discipline on LCA Subsidy

After the GATT Uruguay round, the rules on subsidy that could be applied to the LCA sector were altered in certain key areas. These rules now characterize the subsidy disciplines of the World Trade Organization (WTO), which replaced the GATT in January 1995. The WTO is now the major international regulatory instrument for world trade relations.

A fundamental change in the new WTO regime is the abandoning of the principle of unanimity, which had characterized the previous GATT system. In consequence rulings of the new Disputes Settlement Body (DSB) and the Appellate Body cannot be overturned by the veto of individual states. As well as these general regime changes there also more specific consequences for LCA. The ceilings for indirect and direct supports, which characterized the 1992 EU/U.S. Bilateral, were not adopted in the new code or any of its footnotes. Of special interest is the fact that subsidy for R&D may be covered in the new WTO disciplines.

The basic GATT provision that subsidies must not be used in ways that harm or threaten to harm one's trading partners has been re-enacted in the framework of the World Trade Organisation through the Agreement on Subsidies and Countervailing Measures (ASCM). Accordingly, ASCM, which contains three provisions specific to the civil aircraft sector, now contains the substantive provisions for challenging subsidies before the WTO, the procedural rules being set out in the Dispute Settlements Understanding 1994 (DSU), also one of the key regulatory instruments of the Uruguay Round.

Under ASCM, a subsidy is defined as a benefit-conferring financial contribution by a government (or other public body) involving a transfer of funds, foregoing of government revenue, goods or services provided by a government or payments by or on behalf of a government to a funding mechanism; or any form of benefit-conferring income or price support as described in GATT 1994 Article XVI (which deals with export subsidies), (ASCM Article 1.1).

A subsidy can only be challenged if it is specific to an enterprise or industry or groups of enterprises or industries or to enterprises in a particular geographic region, (ASCM Article 1.2 and Article 2). ASCM also distinguishes between three categories of subsidy: *red light* (actionable per se) *yellow light* (actionable if injury suffered) and *green light* (actionable only if resulting in serious adverse effects causing damage which is difficult to repair), (ASCM Parts II, III and IV).

Legal Rules Applicable

The key elements of the ASCM definition of subsidy are:

1. A financial contribution by a government
2. A contribution which confers a benefit
3. A benefit which is specific, i .e. limited to certain enterprises.

General regional aid is not therefore covered by Article 2, provided it meets certain criteria laid down in Article 8 (2)(b) which deals with non-actionable subsidies. Note that a subsidy may be actionable if limited to certain enterprises located within a designated geographic region.

Prohibited Subsidies (actionable per se) – Red Light

ASCM contains an outright prohibition on only one category of subsidy: export subsidies i.e. subsidies contingent upon export performance or export substitution by favouring local products, (ASCM Article 4).

Actionable Subsidies (injury requirement) – Yellow Light

ASCM Article 5 provides that no member should cause, through subsidies, adverse effects to the interests of another Member, i.e.

1. Injury to the domestic industry of another member
2. Nullification or impairment of benefits accruing from GATT 1994
3. Serious prejudice to the interests of another member.

Article 6, itself entitled *serious prejudice,* creates a presumption of serious prejudice, notably in the case of ad valorem subsidization of a product exceeding five percent, direct forgiveness of debt, defined as forgiveness of government-held debt, and grants to cover debt repayment. These presumptions do not, however apply to the aircraft industry because of two exceptions, which are described in the next section, (ASCM, Article 3, article 6 (1)(a) and (6)(1)(d)).

The serious prejudice presumption was a matter of considerable concern for the LCA industry in the closing phase of the Uruguay Round. Negotiators secured exceptions for the aircraft industry from the two paragraphs enumerated above, but they operate in different ways. A footnote to Art.6 (1)(a) states that: 'Since it is anticipated that civil aircraft will be subject to specific multilateral rules, the [five percent] threshold in this subparagraph does not apply to civil aircraft'. A footnote to Art.6 (1)(d) states: 'that Members recognize that where royalty-based financing for a civil aircraft programme is not being fully repaid due to the level of actual sales falling below the level of forecast sales, this does not of itself constitute serious prejudice for the purposes of this paragraph'. This reflects the cyclical nature of civil aircraft sales. It was inserted to ensure that where governments allow leverage in repayment terms of a result of a market downturn, that would not constitute a presumption of serious prejudice.

Serious prejudice may arise in four enumerated situations where the subsidy leads to one or more of the following, (Article 6 (3)).

(a) displacement or impediment of imports of a like product
(b) as (a), but from a third country
(c) price undercutting
(d) an increase in world market share.

Articles 6, (4) and (5) introduce important clarifications to Articles 6 (3)(c) & (d), notably by requiring comparisons to be made with a non-subsidized like product. In practice any complaint will bring under scrutiny the products of the complaining party and any subsidisation by that party will be taken into account. This is particularly relevant to the civil aircraft sector as any complaint by one LCA company would be unsustainable if the competing product of the rival corporation had also been subsidized – even if not to the same proportions.

Non-actionable Subsidies –Green Light

ASCM Article 8 identifies as non-actionable subsidies assistance for research activities conducted by firms or by higher education or research establishments on a contract basis with firms if the assistance covers not more than 75 percent of the costs of industrial research or 50 percent of the costs of pre-competitive development activity and provided the assistance is limited exclusively to certain costs.

Article 8 (2)(a) is, however, subject to a footnote which states that: 'Since it is anticipated that civil aircraft will be subject to specific multilateral rules, the provisions of this subparagraph do not apply to that product'. In other words, subsidies for R&D for civil aircraft are actionable under ASCM Part III, notwithstanding the generality of Art.8 (2)(a). This is central to the focus of our analysis of U.S. LCA, as this provision applies to NASA and DoD R&D funding, regardless of the percentage it represents of the overall costs of the particular research project.

The U.S. has also expressed concern about state aid granted by member states and approved by the European Commission under the Framework for aids for R&D. Indeed, in 1999 under this heading Honeywell mounted a European Court challenge to the Commission's approval of launch aid to Sextant. However, the fact that the action was promptly dropped may have resulted from the realization that the EU could reciprocate and launch a case against a U.S. supplier company. This is of interest because U.S. responses to earlier reports on LCA subsidy have insisted that only R&D contracts going to LCA prime companies are relevant to the issue of LCA supports. However, with programs such as ManTech, this is certainly not the case.

The ASCM Procedure

ASCM establishes a twin track system for dealing with subsidies as defined in ASCM Part I. Under Track 1 (ASCM Parts II & III), a member wishing to challenge a prohibited or an actionable subsidy under ASCM may trigger consultations. If consultations are not successful, a party may call for the establishment of a panel. Unless there is an appeal, the panel's report normally forms the basis for the decision of the WTO Disputes Settlement Body (DSB). If the panel report is appealed it goes instead to the Appellate Body for a decision. In either case there will be a time limit for compliance. Failure to comply may lead to countermeasures and or to arbitration. Disputes under Track 1 are subject to the WTO Disputes Settlement Understanding (DSU). However, certain time limits are shorter for disputes arising in respect of actionable subsidies and shorter still in respect of prohibited subsidies. Under Track 2 (ASCM Part V), members are alternatively entitled to impose countervailing duties, in accordance with Article VI of GATT 1994 and the terms of the ASCM. Track 2 provides for consultations with members whose products may be subject to investigation. Essentially, the Track 2 procedure is a domestic procedure. However, failure to comply with the WTO rules in the domestic procedure may itself give rise to a WTO challenge under the DSU. Each of Parts II, III and IV involve slightly differing procedures for challenging subsidies depending on their nature.

Conclusion

In this chapter we have outlined the history of the documentation of LCA sector subsidy in order to give a context to the analysis of U.S. federal subsidy in this report. We have included an overview of the changing regulatory framework in order to show how the disciplines which apply to subsidy in the LCA sector have become, both more precise and more binding on individual states. From the European perspective the ASCM procedure is significant because we believe that the U.S. government is liable under this new discipline with respect to its R&D support for the LCA industry, via the NASA and DoD programs we document.

3 U.S. Industrial Policy for the Aerospace Industry

> In return for prosperity, American society accepted the legitimacy and permanence of the core American Corporation… American officials took as one of their primary responsibilities the continued profitability of [these] corporations, (Reich, 1991, p. 58).

The U.S. Industrial Policy for the Aerospace Industry

This chapter provides an overview of the U.S. aerospace industry, of which the LCA sector is a subdivision. Its purpose is to show that there is a U.S. industrial policy for aerospace that significantly benefits Boeing, the one remaining U.S. LCA manufacturer. The wider aim of the chapter is to place our study of the U.S. LCA sector in a broader industrial context.

In the analysis below we will show that the aerospace industry makes the following positive contributions to the U.S. economy.

- ➢ More than 800,000 well-paid/high value-added jobs
- ➢ The largest manufacturing surplus in the U.S. balance of trade
- ➢ Technological leadership which frequently spins-off into other industrial sectors
- ➢ Contracts which support business activity in a vast number of small and medium-sized enterprises across the U.S.
- ➢ The technological basis for national security
- ➢ International status and prestige.

The quotation above, from former U.S. Secretary of Labor, Robert Reich, is an apt introduction to this chapter on the American aerospace industry. In the continuing debate about federal subsidy of U.S. aircraft manufacturers a key problem is a fixation in the United States on free market principles. Quite simply U.S. economic ideology places a blanket of confusion over the real nature of federal industrial policy to industries such as aerospace. The problem is not new.

In the 1970s, leading U.S. industrial economist John Kenneth Galbraith observed that:

> Only someone with an instinct for inconvenience suggests that firms such as Lockheed or General Dynamics, which do most of their business with the government, make extensive use of plants owned by government, have their working capital supplied by the government, have their cost overruns socialized by government... are anything but the purest manifestations of private enterprise, (Galbraith, 1973, pp. 3-4).

Bearing in mind the points made by Reich and Galbraith we aim to show in this chapter the nature of U.S. industrial policy for the aerospace industry and its relevance to the LCA sector.

In our view, the more overt industrial policy in aerospace of recent years is a response to European competition and the huge job losses experienced in the U.S. in the late 1980s and early 1990s. Of special significance is the fact that during the tenure of the Clinton Administration U.S. LCA manufacturers have received increasing levels of support through the Clinton Aviation Initiative and other policy instruments implemented through NASA and the DoD.

The U.S. Aerospace Industry: an Overview

> America's aerospace industrial base is crucial to our nation's health. It's crucial because of national security. It's crucial to our continued technological leadership... it's crucial because of jobs, (Douglas, 1998, p. 1).

As the quotation above illustrates the U.S. aerospace industry is in the first rank of strategic importance. The industry, which in 1998 generated sales in excess of $140 billion, (Aerospace Industries Association, 1998) plays a critically important role in preserving and enhancing American economic, political and military power in the global system and is one of the most prominent symbols of America's technological and market dominance. It is therefore hardly surprising that the U.S. aircraft manufacturing industry has 'Throughout much of its history... benefited from a makeshift but nonetheless effective industrial policy', (Tyson, 1992, p 157).

Since 1993, the industry has pursued rapid and far-reaching consolidation and rationalization strategies to overcome the twin problems of global economic recession and post-Cold War defence budget reductions.

Between 1990 and 1995, sales of U.S. commercial aircraft fell by some 37%, while military aircraft sales decreased by about 20%. In the 1990s, as former Presidential advisor Laura Tyson has asserted, the industry confronted a dual challenge. On the one hand it was faced by declining defence budgets, on the other it was posed with the compelling challenge of Airbus Industrie, (Tyson, 1992, p. 155).

Industrial Restructuring

In the U.S., the integration of civil and military aerospace interests, under-pinned by extensive and deep-rooted government research and development programs, has been undertaken with remarkable speed in the last six years. Between 1992 and 1997, 32 defence companies, principally in the aerospace sector, were concentrated into just nine, with the loss of about one million jobs. As a result of this unprecedented wave of aerospace industry mergers, the major U.S. aerospace companies dominate global aerospace markets in terms of sales revenue. Table 3.1 below shows the 10 largest post-merger aerospace companies in the world in 1999 by sales, with U.S. companies taking eight of the top positions and Boeing and Lockheed Martin holding the top two positions in the list.

Table 3.1   The Post-Merger Aerospace Ranking by Sales (U.S. $bn): 1999

Rank	Country	Company	U.S.$bn
1	USA	Boeing	55.4
2	USA	Lockheed Martin	26
3	EU	EADS (Includes CASA)	21.8
4	GB	BAE Systems	20.5
5	USA	Raytheon	17.5
6	USA	United Technologies	12
7	USA	General Electric	10.3
8	USA	Honeywell	9.8
9	USA	Northrop Grumman	9.1
10	USA	TRW	5.9

(Source; IPG/*Flight International,* 8-14, September 1999, *Flight International,* 20-26 October, 1999).

In 1996, the announcement of the impending merger of Boeing, an aerospace giant which already dominated the commercial sector, and McDonnell Douglas sharply raised the profile of military aerospace activities within the Boeing organization, and with it the prospect of greatly enhanced opportunities for military/civil synergies and technology transfers. Having already acquired the defence and space business of Rockwell International for $3.2 billion, the merger with McDonnell Douglas increased the share taken by military aerospace to about 40% of Boeing's $45 billion revenue for 1997. After the merger was completed Boeing acquired MDC's $13bn civil and military business, including $7bn worth of military contracts. With projected sales in 1999 of $55bn Boeing has emerged as the real colossus of global aerospace and is now able to benefit more than ever from defence/commercial synergy (see chapter 1) and contrasting defence/civil cycles. In 1997 the merger bequeathed to Boeing the following:

- 84% of all LCA in service (88% including freighters and military variants)
- 60% of current sales of LCA
- 70% of LCA backlog
- Monopoly of both 100 seater and >350 seat transports.

(Aerospace Strategy Research Centre Analysis)

This merger is an example of horizontal integration within the industry, which as we have seen, has created a global giant. McDonnell Douglas, for example, manufactured the F-15 and F/A-18 fighter aircraft, the C-17 transport, the Apache gunship, and (with British Aerospace), the Harrier jump-jet and Goshawk trainer. Boeing, on the other hand, while predominantly a civil aerospace manufacturer, has a substantial share of the advanced F-22 fighter aircraft, the E3 AWACS early-warning aircraft, the Chinook and Comanche helicopters and the Osprey tilt-rotor aircraft and, with Lockheed Martin, won the contract to build one of the two prototype Joint Strike Fighters, which, for the winning contractor will be a program worth potentially $100bn, (Financial Times, September 3, 1998).

Hence, a company such as Boeing, with a large domestic market for its aircraft and a "family" of products designed to meet the needs of its commercial airline and military customers, has the potential to extract both scale and scope economies, significantly reducing average costs of production. In consequence it should be able to secure a key competitive advantage over rivals in the global marketplace.

U.S. LCA Research and Development Consolidation

As we showed in chapter one substantial R&D funding is fundamental to maintaining a competitive edge in global aerospace. With regard to government funded support and subsidy the 1997 merger with McDonnell Douglas has allowed Boeing to absorb the DoD and NASA R&D, which previously accrued to MDC. As Interavia's Oliver Sutton notes, 'Boeing's... merger with McDonnell Douglas will further facilitate access to dual-use technology R&D funded by the U.S. defence department and NASA, (Interavia, May, 1998, p. 18). In particular Boeing can now reap the benefits of synergies in defence/civil aerospace through its inheritance of McDonnell Douglas's federal R&D contracts. As table 3.2 below reveals the volume of this federal spending is enormous, with the merged company accruing billions of dollars of former MDC contracts.

Table 3.2 Boeing and MDC Federal R&D Contracts 1992-1996, Excluding Procurement (Current U.S. $000s)

Year	1992	1993	1994	1995	1996
Boeing	512,645	968,450	1,579,109	1,970,642	1,536,985
MDC	2,102,518	2,035,136	2,006,177	2,153,732	1,998,266
Total	2,615,163	3,003,586	3,585,286	4,124,374	3,535,251

(Source: Government Executive 1992-1996)

In the U.S., the volume of spending for R&D cited above is not seen as financial support or subsidy for the commercial aerospace industry. But this is not a plausible conclusion. Without question, the giant corporations leading the aerospace industry in the U.S., particularly the large civil aircraft sector, are critically important to present and future American economic well-being. The R&D support shown in table 2.2 is thus a government investment to ensure that the industry continues to generate positive externalities for the U.S. economy. As we write the American Aerospace Industries Association (AAIA) is lobbying hard for a Presidential enquiry into U.S. aerospace and is seeking even more federal R&D funding.

High Skill/High Value Added Employment

The LCA industry in the U.S. plays a critical role in maintaining a healthy domestic economy. Aircraft manufacture has long made major contributions to the U.S. economy in the form of high-wage and high-skill employment. With the exception of the automobile industry, no other industrial sector contributes as many jobs to the U.S. economy. In 1997, over 850,000 people were employed in the industry, of which almost 300,000 were production workers. While this represents an increase of 40,000 production jobs in U.S. aerospace since 1995, the total is considerably lower than in the peak years of the late 1980's, when overall aerospace employment in the U.S. stood at 1,314,000, with 432,000 production jobs. It has been estimated that each $1 billion of aircraft shipments by U.S. companies maintains or creates 35,000 jobs in the industry (Cantor, 1992).

Aerospace is a knowledge-intensive industry, which employs top scientists, engineers and a vast range of highly skilled employees. The high skill base and knowledge intensity of aerospace makes the aircraft industry "strategic", since it generates "excess rents" or higher returns to the factors of production than they could expect to receive if employed elsewhere in the economy, (Tyson, 1992, p. 155).

A Positive Balance of Trade for the U.S.

The industry also makes a critically important contribution to the U.S. balance of trade position, all the more important given the large trade deficit that has proved problematic for the U.S. for so long. Indeed, since the late 1950's, aerospace has been the principal industrial contributor to U.S. export revenue. It exports more manufactured goods in terms of value than any other U.S. industrial sector and, with its dominant position in global aerospace markets, has been able to generate a larger trade surplus than any other sector of U.S. industry. In 1995, the U.S. aircraft industry generated a trade surplus in excess of $21 billion (Napier, 1996). In 1998, U.S. aerospace exports and the industry trade surplus reached record levels. U.S. aerospace exports increased dramatically by $14 billion to an aggregate of $64 billion and the aerospace trade surplus expanded to a record $41 billion, an increase of $8.7bn over the 1997 figure.

Table 3.3   U.S. Aerospace Trade Surplus

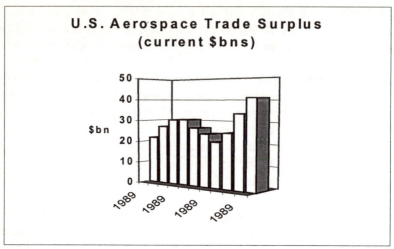

(Source, Office of Aerospace, Department of Commerce)

The most important contributor to aerospace export growth in 1998 was the civil aircraft sector where exports increased by $8 billion to reach $29 billion (AAIA, 1999). As table 3.3 above shows the surplus in aerospace trade has been a consistent feature of the U.S. economy over the last decade with strong growth in the last two years.

Technological Leadership

The role played by the U.S. aircraft manufacturers in technological leadership owes much to its unique relationship with the U.S. government, particularly through the research support it receives from NASA and the DoD. On the Boeing B-777 alone, described as being a 'landmark in commercial aircraft development by virtue of its significantly advanced performance, efficiency, safety and environmental acceptability', (National Aeronautics Association, 1996), the following enhancements derived from NASA funding:

  ➢ transonic supercritical airfoils
  ➢ advanced metallic alloys
  ➢ composite materials
  ➢ digital flight controls
  ➢ glass cockpit instrument displays

> ➤ flight management systems
> ➤ laminar flow control concepts
> ➤ fatigue and fracture methodology.

(For more detail see chapter four).

Of critical importance to the preservation and enhancement of U.S. competitive edge in global markets, the U.S. aerospace industry also plays an important role in extending leading-edge technologies to other critical high-technology industries through the synergies that result from aerospace technological advance. Innovations in aerospace technology, particularly in the large civil aircraft industry, have a positive effect on those core technologies that stimulate the development of many new products as well as impacting significantly upon production processes in many other sectors of the U.S. economy (Van Tulder and Junne, 1988). Remarkably, a Congressional Research Service study found that of some 429 identified sectors of the U.S. economy, 340 were interrelated with aerospace and 150 supplied products or services directly to the aircraft industry, (Cantor, 1992, p. 43).

The tendency to continually pursue leading-edge technological advances helps to ensure that a strong knowledge base is preserved and enhanced. Another important factor is the need for the industry to utilize new innovations rapidly and effectively. This is because the first aircraft manufacturer to introduce into service a new aircraft with a significant technological advance over competitors gains formidable and lasting competitive edge in the market. Early application of new technology in this way, given the inter-locking nature of much of U.S. high-technology industry, means that the diffusion of knowledge through the rest of U.S. industry tends to be accelerated with consequent decreased costs and, most critically, a pronounced sharpening of U.S. competitive edge in many global markets (Tyson, 1988, p. 112).

National Security

The view that the U.S. has an industrial policy for aerospace is doubtless controversial. However, the simple truth is that the industry arose on the back of the U.S's post-1945 requirement for a global defence capability based on aerospace technology. In the post-1945 era, the U.S. State has had, in effect, a covert industrial policy in the aerospace sector. The policy was funded by the

39

Pentagon, its aim derived from security strategy, its beneficiaries the large manufacturers of defence and defence-related equipment and its consequence the maintenance of huge corporations in high technology sectors such as aerospace. As U.S. aviation historian David Thornton notes, 'Indeed, the economic and organizational impact of these capital outlay and procurement policies was so great that the U.S. government had a de facto industrial policy towards the aerospace sector', (Thornton, 1995, p. 25). The linkage between LCA production and security continues today. The LCA sector can be considered to be of strategic importance to the U.S. in the conventional military sense, because of spill-overs between commercial and military operations and the potential for economies of scale and scope, as well as leading-edge technology to be transferred from one sector to the other. In the U.S. the one remaining LCA manufacturer is also a military aircraft producer and thus can draw upon a common pool of expertise, sub-contractors and component suppliers. The shared knowledge and production base of commercial and military aerospace renders the industry strategic from the point of view of national security, since it forms such a crucial element of the military global reach of the U.S. Commercial design and production teams have developed technological innovations relevant to military requirements and vice versa. Sharing knowledge, personnel and production facilities enables the U.S. to maintain the world's leading military industrial base at lower cost and provides enormous scope for wartime surge capacity in production (Lopez and Yeager, 1987, p. 42). Without this crucial symbiosis, the maintenance of a large independent U.S. military industrial complex in the aerospace sector would be prohibitively expensive.

The synergies between the military's emphasis on performance and flexibility and the commercial sector's emphasis on cost and reliability have become increasingly important to aircraft technology and innovation. A globally competitive commercial aircraft industry therefore contributes to a nation's military capability. Both civil and military aerospace sectors also experience recurrent business cycles which are rarely synchronized, thereby enabling employment and production to be maintained by switching employees and resources between declining and expanding aerospace sectors. As a result, production capacity is almost always available when required, enabling the U.S. to offer solid guarantees regarding its military commitments in uncertain situations. As a Congressional Office of Technology Assessment study notes, 'Many companies cite the counter cyclical effect of military business as crucial to their ability to maintain a level workforce and keep design teams together', (COTA, 1991, p. 55). Regarding the question of subsidy, combined military and civil production allows the U.S. government the option of implementing a covert form of industrial policy. As the DoD/NASA Aerospace Knowledge Diffusion Project acknowledges:

> The connection between government procurement and corporate pursuit of civil projects is reflected in the Clinton Administration's decision to purchase 80 McDonnell Douglas C-17s in addition to the existing procurement of 40: the infusion of an additional $17 billion allowed McDonnell Douglas to remain afloat financially and hire over 2000 additional workers, (1997, p. 55).

Recent Federal Policy for Aeronautics

In the period since the 1992 EU/USA Bilateral Agreement on Trade in Large Civil Aircraft, a number of important developments illustrate clearly the degree to which the U.S. Government, directly and indirectly, has enhanced its support mechanisms for the U.S. civil aircraft industry and its key manufacturers. The strategy involves two inter-related elements:

1. Clinton Administration efforts to bolster U.S. civilian high technology industry
2. Dual-Use and Reforms in the DoD acquisition process.

Commercial Pay-off for R&D

Despite some retrenchment in the NASA budget in the mid-1990s, the focus of NASA's activities appears to have shifted towards R&T designed to support aerospace clients. This includes a much greater emphasis now being given to applied research and a commitment to share research findings at an early stage with representatives of the U.S. LCA sector and its suppliers. Furthermore, research programs appear more focused on technological developments which have obvious commercial relevance, particularly research that U.S. corporations are unwilling or unable to finance themselves, (NASA FY 1996 Appropriations Hearing.104th Congress, 1st Session, 1995). Indeed, such was the scale of NASA support for commercially-orientated projects such as the High Speed Research Program (HSR) and Advanced Subsonic Technology Program (AST) that the Congressional Budget Office expressed concern in 1995, noting that: 'the benefits from the R&D supported by the NASA programs in question fall almost exclusively to aircraft manufacturers, their suppliers and airlines', (United States Congressional

41

Budget Office, Reducing the Deficit: Spending and Revenue options, 152, February. 1995). The support under these programs has been considerable for the "industry-active" partners in the Advanced Subsonic Transport program (such as Boeing and McDonnell Douglas). These companies have been the recipients of significant AST R&T funding, for example with a contract for a new aircraft wing, which received $62 million from NASA, (*Aviation Week and Space Technology*, August, 17, 1994, p. 277).

Two trends are worth noting:

1. Greater co-ordination has been encouraged by the U.S. government between federal R&D laboratories and private industry with both "partners" gaining from enhanced funding and more beneficial technology transfer regulations. The new U.S. approach to regaining its former dominance of the global civil aircraft market was formalized in August, 1995 with the publication by the President's National Science and Technology Council report: Goals for a National Partnership in Aeronautics Research and Technology.

2. Since 1995, President Clinton and his trade and commerce officials have undertaken a number of initiatives to promote overseas sales of U.S. commercial aircraft, providing significant support through the U.S. Export-Import Bank.

On October 28, 1993, President Clinton spelt out the key role of technology policy in the new administration, outlining an array of devices to be employed in the future to support U.S. aerospace producers and ensure their competitiveness in the global market. These included:

➢ tax incentives for investment in R&D
➢ switching of federal resources towards basic research and civilian technology
➢ the promotion of defence conversion
➢ direct expansion of federal investment in basic research, through agencies such as the National Science Foundation
➢ the strengthening of collaboration with industry through consortia.

Dual-use Industrial Policy for U.S. Aerospace

In the 1990s the U.S. has made substantial cuts in its defence procurement budget. In order to get better value for money the Pentagon has been seeking more commercial off-the-shelf (COTS) components to incorporate into defence systems. Since 1992 the U.S. Department of Defense has publicly encouraged the development of a dual-use R&D policy, stressing its crucial role in strengthening the commercial sector of the economy, (National Security Council Office of Science and Technology Policy, 1995). As part of this process, the Department has started to reduce the number of precise and restrictive military specifications which manufacturers are obliged to meet. Furthermore, in the U.S., much attention has been given to identifying the critical technologies necessary to maintain both military and economic power. These technologies include:

- ➢ materials synthesis and processing
- ➢ electronic and photonic materials
- ➢ ceramics and composites
- ➢ computer-integrated flexible manufacturing
- ➢ systems management technologies
- ➢ micro and nano-fabrication
- ➢ software, microelectronics and opto-electronics
- ➢ high-performance computing and networking
- ➢ high-definition imaging and displays
- ➢ sensors and signal processing
- ➢ data storage and computer simulation
- ➢ applied molecular biology and energy technology.

(National Economic Council/National Security Council/ Office of Science and Technology Policy, 1995).

As part of his vision of future U.S. technology policy, President Clinton announced in 1993 a range of newly created or expanded technology support programs which included: the Department of Defense's Technology Reinvestment Project (now the Dual-Use Applications Program); the Department of Commerce's Advanced Technology Program (ATP); and the inter-departmental Co-operative Research and Development Agreements (CRADAs).

The purpose of ATP is made clear in a statement on the DOC's web page:

> The goal of the ATP is to benefit the U.S. economy by cost-sharing research with industry to foster new innovative technologies. The ATP invests in risky, challenging technologies that have the potential for a big-payoff for the nation's economy, (Department of Commerce web page).

The critical point here is the long-term nature and multi-sectoral applicability of federal support for R&D in manufacturing technology under such schemes, a distinctive characteristic, which is not replicated in Europe. With dual-use initiatives the linkage between civil and military applications in aerospace is explicitly recognized. Thus while official pronouncements indicate otherwise dual-use is clearly a means of using military funding to enhance civil technology. As a National Science and Technology report argues:

> The significant basic technological commonality between military and civil aviation products and services must be exploited to increase the productivity and efficiency of our R&T development activities. This requires government and industry, working together, to actively seek technological goals that are common to both civil and military applications, (National Science and Technology Council, 1995, p.5).

Conclusion

In the aerospace industry, as in many other major sectors of modern economies, it must be recognized that governments are now deeply involved in strategies aimed to secure global market success. In the case of the United States, the aerospace industry - particularly the large civil aircraft sector - is a powerful economic force, driving technological change to its most advanced degree and helping to transfer such gains to other high-technology sectors, generating highly-skilled employment opportunities. As we have seen the industry is also a star performer in trade terms, offsetting U.S. weakness elsewhere. Aerospace has also fuelled the expansion of a vast number of smaller supply companies across the country, with significant positive regional economic effects. Finally, the industry is pivotal for national security. Thus not surprisingly, given the strategic importance of aerospace, the U.S. government deploys a number of policy instruments to support its LCA industry.

In this chapter we have sought to spell out the nature of U.S. public policy for aerospace and its implications for the American LCA sector. Given the official statements made concerning U.S. policy for aeronautics and the paramount strategic significance of the industry it is difficult to believe that the U.S. LCA sector is not the beneficiary of federal financial support. In chapters four and five we will spell out in more detail the funding and programs which we contend represent federal subsidy.

4 NASA Subsidy of R&T for the U.S. Large Commercial Aircraft Sector

This industry, principally the LCA sector, is unique in that it has been the beneficiary of federally funded research and development (R&D) for nearly a century, (T. Pinelli, et.al., 1997, p. XIII).

NASA's aeronautics R&D program helps U.S. aircraft manufacturers develop and adopt new technologies in two ways: by conducting research in-house and transferring the results to companies, and by contracting with companies to perform specific research tasks, usually in co-operation with in-house NASA research. Further, NASA researchers themselves act as a sort of free consulting service for industry engineers having trouble with technical problems. The availability of technologies developed and tested at NASA's expense and risk helps aircraft manufacturers incorporate new capabilities into their products at diminished cost or risk, just as military developments do. Often these technological advances result in gains in competitiveness for the firms that use them... (Congressional Office of Technological Assessment, 1991, p. 73).

Overview

In this chapter we aim to spell out in detail the benefits accruing to the U.S. large commercial aircraft (LCA) industry from NASA Research and Technology (R&T) Programs. This involves four distinct steps. First, we will analyze NASA's mission and the fundamental role that the agency plays in disseminating aeronautics knowledge to U.S. companies. Secondly, we will outline the specific NASA programs and their subdivisions. Thirdly, we will indicate the scale of the NASA budget and its subdivisions. Finally, we will provide estimates of the financial benefit of NASA programs for companies in the U.S. LCA sector. In making this assessment we will use "top down" figures from the U.S. Congress Conference Committee budget authorization, the NASA Office of

Aeronautics and the American Association of Aerospace Industries (AIA). But these estimates will be supplemented with "bottom up" data taken from NASA R&T contracts, obtained under the Freedom of Information Act. In our view the latter approach is very useful in tracking the flow of benefits to individual companies in the U.S. LCA sector. In previous studies of U.S. LCA subsidy the precise benefit of NASA programs to the LCA sector has been difficult to determine because of lack of contract data. Accordingly, we have sought to provide this information in our NASA analysis. For a full account of the methodology used in our analysis the reader should refer to Appendix A.

NASA's Mission and Function

In the following section we detail the role and function of NASA and its predecessor NACA. The National Aeronautics and Space Act founded NASA in 1958. In its charter NASA has the following three objectives which relate to aeronautics:

1. The improvement of the usefulness, performance, speed, safety, and efficiency of aeronautical vehicles;
2. The establishment of long-range studies of the potential benefits to be gained from aeronautical activities;
3. The preservation of U.S. leadership in aeronautical technology, (National Aeronautical and Space Act, 1958, P.L 85-568).

The third strategic objective cited above makes clear the imperative at NASA to reinforce the technological competitiveness of the U.S. large commercial aircraft sector. Seen in the abstract, the implication of this aspect of NASA's mission may not be clear. But in concrete terms the NASA mission means a close working relationship with U.S. companies. According to Golich and Pinelli, aeronautics leadership has been achieved by the following strategic orientation:

> U.S. aeronautical leadership was obtained by close co-operation between state and industry, following a "mission oriented" strategy characterized by large- scale project work centering on firms with a heavy emphasis on areas such as defense. Extensive federal support for production, transfer, and use of aeronautical knowledge and technology began in 1915 under

the auspices of the National Advisory Committee for Aeronautics (NACA), (Golich and Pinelli, 1998, p. 7).

In the long-standing contention over whether the U.S. government subsidizes the American LCA industry much has turned on the role of NASA. Here we seek to show that NASA does provide a subsidy to the U.S. LCA sector by outlining the function that the agency plays in promoting the competitiveness of U.S. civil aeronautics. In so doing we have been amply assisted by the *DoD/NASA Aerospace Knowledge Diffusion Research Project,* which was published as a two-volume study in 1997, (Pinelli et.al, 1997). As we will show this study demonstrates clearly the purpose of public funding for the U.S. large commercial aircraft industry. The report also reveals a fundamental fact about federal funding, which has obscured the role of NASA and the DoD and has undoubtedly led to confusion over the subsidy question. As the authors assert:

> Although millions of tax-payers dollars are spent annually for this purpose little is known about how the knowledge and technology resulting from federally funded aerospace R&D diffuse at the individual, organizational, national and international levels, (Pinelli, et.al. 1997, p. XIII).

As we have seen NASA's mission dates back to 1915 and the creation of its predecessor, NACA. NACA's original mission was to further research 'into the problems of flight with a view to their practical solution', (COTA, 1991, pp. 65-66). Although originally not intended to conduct its own research NACA was soon compelled to build its own facilities in California, Ohio and Virginia, because private industry was not producing the innovations government demanded. Today these facilities, dating back to the 1920s, 1930s and 1940s, constitute NASA's Langley, Ames, Lewis (now Glenn) and Dryden research centres.

Utilizing the now famous research centres NACA made a number of pioneering breakthroughs which contributed to resolving the problem of aerodynamic drag, which had limited further gains in speed and fuel economy for aircraft in the inter-war years. Particularly significant were the development of laminar flow airfoils and the subsequent creation and data cataloguing of a whole family of airfoils, used to give engineers an off-the-shelf tool for design optimization. Thus NACA research was critical in driving forward U.S. advances in aerodynamics. As a COTA study notes, 'Up to the Second World War, the NACA's facilities and expertise made it the leading force in aeronautical research in the United States', (COTA, 1991, p.67).

After World War Two, NACA became less important in aeronautics R&D as attention turned to the defence aerospace requirements created by the Cold War. But this did not blunt the commercial application of aeronautics R&D, as DoD programs were the genesis for technologies used in the U.S. jetliner industry. Using a very direct policy instrument, the DoD developed prototypes and then passed them on to industry for commercial development. A process it also used in nuclear power and synthetic fuels, (Heppenheimer, 1995, pp. 1-2). In 1958, NACA was transformed into the National Aeronautics and Space Administration (NASA) and with the new accent on space exploration, aeronautics came a poor second in terms of overall R&D funding. But in the late 1960s and early 1970s, the tide turned for NASA and the agency began to pursue the role it now plays in aeronautics R&D. In the 1970s NASA adopted a "proof of concept" philosophy which required that research must go beyond the laboratory in order to demonstrate its viability in action, (COTA, 1991, p. 68).

It is essential to realize that the "proof of concept" philosophy fundamentally changed NASA's relationship to the U.S. LCA industry. As we have seen, a correct account of U.S. federal R&D funding must recognize that the funding fills a gap left by the free market. With "proof of concept" programs this indirect subsidy becomes more directly focused on the commercial aims of the private sector. "Proof of concept" required that NASA adopt demonstrator programs in addition to the famous experimental X-series, which primarily addressed military applications and supersonics.

Demonstration programs take the concepts and results from aeronautical R&T and embeds them in prototypes or working systems. They are bridges between R&T and development. To put it simply demonstrators are a critical step between ideas and tangible real products. Demonstrators provide "proof of concept" and offer preliminary data on likely performance characteristics and reliability. With recent developments in computerized design tools, such as CATIA and CADCAM, physical demonstrators can be integrated with virtual ones in order to speed up the aircraft development cycle. Demonstrator programs are particularly relevant to our discussion here because they represent precisely the form of R&D activity not likely to be funded by private companies.

Figure 4.1 NASA Demonstrators

Selected	NASA Demonstration Programs (Civil and Military) 1970 onwards
B-757	Hybrid laminar flow testbed (subsonic)
B-737	High lift flap technology
F-18	High angle-of-attack research vehicle
SR-71	Flight research testbed for high-speed civil transport
UH-60	RASCAL Helicopter, map of the earth flight guidance, GPS precision navigation flight system
XV-15	Certification profiles for noise abatement, failure modes and handling qualities
F-15	Highly integrated digital electronics control
F16-XL	Hybrid laminar flow control (supersonic) for civil transport
X-29	Thin, supercritical-section forward swept wing & close coupled canard
X-30	National Aero Space Plane (NASP)
X-31	Quasi-tailless configuration to explore possible drag & weight savings in future civil and military designs
X-32	Boeing JSF concept demonstrator
X-33	Venture Star sub-orbital Reusable Launch Vehicle (RLV)
X-34	Orbital Sciences RLV technology demonstrator
X-35	Lockheed Martin JSF concept demonstrator
X-36	Boeing tailless aircraft demonstrator (original contractor was MDC)
X-37	Designation reserved for Future-X low–cost access to space demonstrator
X-38	International Space Station Crew Return Vehicle technology demonstrator
X-39	X-number still to be assigned
X-40A	Differential GPS automatic landing guidance system demonstrator
X-41	Classified program
X-42	Classified program
X-43	Part of the Hyper-X hypersonic experimental vehicle program

(Source: NASA/*Flight International* (6-12/1/99).

In the USA funding of technology demonstrators is a strategic policy and often follows an appeal to Congress for funds on the grounds that the U.S. is falling behind in an essential domain of technology. Here the strategic dimension can be either related to economic well being or issues of national security, (Hansson, 1997, p. 6).

Following NASA's conversion to "proof of concept" R&D, a number of civil and military demonstrator programs were initiated to the benefit of the U.S. LCA sector. NASA demonstrators are detailed in figure 4.1 above. Military demonstrators have been included in the selections above because our analysis is premised on the view that fundamental spin-off potential exists from military to civil programs, (see our discussion in chapter five). It is certainly the case that the DoD is less likely to provide whole aircraft prototypes for commercial conversion, but with 50% of an aircraft's value now in equipment and engines, (ARC estimate), the migration of systems, components and materials to LCA must be a major focus of attention. In recent years new composite materials and new alloys developed in military programs have migrated across to civil transports and have provided cost savings for U.S. LCA programs. Moreover, some of the more revolutionary changes in manufacturing technology have been derived from the defence side of the business. As the COTA notes, 'The development and testing of composite materials on military programs has been very important to their successful introduction on commercial programs', (COTA, 1991, p. 41). With regard to space demonstrators the NASP X30 program assisted in the development of new titanium alloys that Boeing utilized on the nacelles of the B-777.

In the early 1980s NASA's assistance to the U.S. LCA sector was challenged by the Reagan administration, which declared that, 'any technology development.... with relatively near-term commercial applications will be curtailed as an inappropriate federal subsidy', (quoted COTA, 1991, p. 68). However, reports by the Office of Science and Technology Policy and the National Research Council succeeded in changing the administration's view. As a result the threat to the aeronautics component of the FY 1983 budget was removed. Nevertheless official political concern over subsidy resurfaced in 1995 when it was stated that, 'the benefits from the R&D supported by the NASA programs in question fall exclusively to aircraft manufacturers, their customers and airlines', (Congressional Budget Office, 1995).

The New Technology and Trade Agenda

The response to the threatened Reagan budget cuts emphasized the significance of NASA for national security and the need to compensate for underfunding of R&D by private firms, (COTA, 1991, p. 69). However, as the decade wore on the U.S. became increasingly sensitized to balance of trade problems and threats to its leadership in high technology from both Europe and Japan. A defining moment in the transition to a more direct industrial policy came in 1988 with the Omnibus Trade and Competitiveness Act. The preamble to the Act states: 'The highest priority of the United States government shall be to pursue a broad array of domestic and international policies... to guarantee the continued vitality of the technological, industrial and agricultural base of the United States', (Section 1001 (a)(4)). Heaton notes how there was 'a steady, incremental policy shift toward a more co-operative and activist government role in promoting technological development', (Heaton, 1989, p. 87). This policy shift was undoubtedly of paramount importance for aeronautics. Towards the end of the 1980s the U.S. Congress was calling for NASA to be more directly focused on the commercial viability of U.S. LCA programs in order to ward off the challenge in LCA from Airbus Industrie, whose products were widely believed to be at least the technological equal of those of the U.S., (Tyson, 1992, p. 155). A 1987 Senate Committee on Commerce, Science, and Transportation directed NASA to:

> Prepare a multi-year technology development and validation plan that will help the United States retain its leadership in aeronautics research and technology and compete in the international marketplace for future civil aircraft...[and] assure continued U.S. leadership in future civil aircraft markets, (U.S. Congress, Report of the Senate Committee on Commerce, Science, and Transportation on the NASA Authorization Act of 1988, June 24, 1987).

The recommendations outlined above have been underscored during the last decade by a number of reports produced by the National Research Council (NRC), the Office of Technology Policy (OST) and the National Science and Technology Council (NSTC), (Pinelli, et.al, 1997, p. 114). Careful scrutiny of these reports reveals a continuing anxiety in U.S. policy-making circles concerning the threat to U.S. LCA leadership from

Airbus Industrie. What is abundantly clear in these documents is a consensus in the U.S. that aeronautics R&D must be more closely linked to American industrial and commercial interests. In the past, policies to support the interests of one U.S. LCA manufacturer have been problematic for U.S. policy makers because this would appear to be favouritism. However, with the merger of Boeing and McDonnell Douglas in 1997, the U.S. LCA business is now in the hands of a national monopoly. Although, even before 1997, it was clear that the U.S. LCA business was largely concentrated in the Boeing Corporation, because of the commercial failure of McDonnell Douglas's most recent products, such as the MD11. With the merger, one of the major internal impediments to supporting the U.S. LCA sector with publicly funded RT&D has now been removed.

Aeronautical Research and Technology Program and Budget Structure

Figure 4.2 NASA's Aeronautics Budget Subdivisions

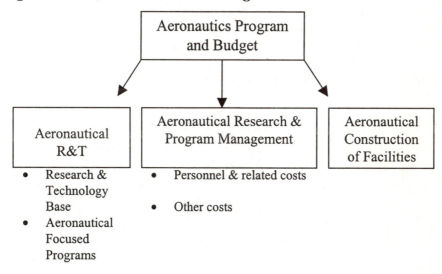

In this section we outline our conception of the NASA budget and program structure. The aim of our analysis is to isolate the actual financial benefit that goes from federal funding to the U.S. large commercial aircraft industry. In consequence we use a method that identifies civil and dual-use programs as distinct from military-only R&T. This discrimination was achieved by a line by line evaluation of NASA's R&T program areas. The reader should note that budget figures given in this section come from NASA's Office of Aeronautics budget statements. The figures differ slightly from the data given below, because those data represent our estimation of real expenditure after we have applied the formula outlined in Appendix A. Other differences arise because NASA's own statistics do not give specific figures and costings for the Research and Program Management aspect of the Aeronautical Research and Technology budget.

Of the three areas of the Aeronautical Budget, the key one is clearly Aeronautical Research and Technology. Here funding is provided for major R&T programs from which technology is transferred to the U.S. large commercial aircraft sector. The R&T Programs are divided into two broad areas of activity. The R&T Base Program, which sponsors and funds long-range, generic aeronautical research and the Aeronautics Focused Programs, where research is geared to more specific near term goals. The two major focused programs are the Advanced Subsonic Transport (AST) and the High Speed Research Program (HSR). Budget outlays for these are given below in table 4.1. As we will indicate below in the section on the AST Program NASA R&T has already generated new wing technology that could be utilized in current aircraft programs.

Table 4.1   **NASA R&T Program Budget**

(Numbers in U.S. $ mn)

	FY93	FY94	FY95	FY96	FY97	*FY98*
R&T Base	436.5	448.3	366.3	354.7	404.2	*418.3*
AST	12.4	101.3	150.1	169.8	173.6	*211.1*
HSR	117.0	187.2	221.3	233.3	243.1	*245.0*
Other	299.7	83.9	144.3	159.5	23.3	*45.7*
TOTAL	*865.6*	*1,020.7*	*882.0*	*917.3*	*844.2*	*920.1*

(Source: NASA Budget, 1998).

Regarding these budget figures, one can see clearly the impact of the 1994 Clinton Aviation initiative, which enacted the recommendations of the National Research Council and Office of Science and Technology reports cited above. These initiatives were also a component of a new industrial policy. In February 1993 President Clinton unveiled a $16.5bn national investment plan aimed at improving the national civil technology base. Included in the plan were proposals to make significant improvements in the aeronautics research infrastructure in the USA, (*Aviation Week and Space Technology,* 1 March, 1993, pp.18-19). In parallel, NASA's FY 1994 budget proposal saw a sharp increase for aeronautics to $1.55bn, up $450mn on the previous year. The budget proposal also included significant increases for the two major Focused Programs, the High Speed Research Program (HSR) and Advanced Subsonic Transport Program (AST). This can hardly be a coincidence, for as a series of national executive level reports continued to emphasize the competitive challenge to U.S. LCA from Airbus, one sees a corresponding new focus at NASA on commercial programs, with a proposal for large spending increases on LCA related research. Two aspects of the pattern of funding are worthy of comment. Firstly, the sharp increase at the time of the Clinton Aviation initiative and secondly, the continual increases through the 1990s in funding of the Focused Programs. In 1994 the Clinton Aviation Initiative proposed an 18% increase in NASA aeronautics funding and the impact of that is clearly apparent in the data above. The reader should note that these figures do not include additional NASA spending on Research and Program Management (RPM) and Construction of Facilities, which are detailed later in this chapter.

Regarding the key question of whether the NASA funding amounts to subsidy for U.S. LCA programs, the following statement of U.S. Transportation Secretary, Frederico Pena, at the time of the launch of the Aviation Initiative is highly pertinent:

> ...While eschewing any return to regulation, we have defined a new role for government as an active player in aviation. One example of this philosophy in the Administration's initiative is the proposal to increase NASA's budget by 18%, so that the agency can subsidize launches for private-sector projects. The programs to receive the bulk of this funding are the Advanced Subsonic Program and the High Speed Research Program ... (Statement of U.S. Transportation Secretary, Federico Pena, Washington D.C., 6 January 1994).

NASA Research & Technology Programs

The following section details the overall NASA R&T programs and their specific direct budgets. The quotations below indicate the commercial relevance of these programs and their function.

> NASA carries out its aeronautics mission in close partnership with the DoD, FAA, U.S. industry and academia. The program reflects the continued need to address critical technology and performance barriers and to strengthen technology development in selected high-payoff areas vital to our long-term leadership in aviation, (NASA Office of Aeronautics).

> The National Aeronautics and Space Administration (NASA) funds the development of technology and systems intended for use in commercial airliners – both subsonic and supersonic – with the explicit objective of preserving the U.S. share of the current and future world airliner market, (Congress of United States Budget Office, Reducing the Deficit: Spending and Revenue Options, A Report to the Senate and House Committees on the Budget, 152, February 1995).

R&T Base Program

In order to outline the exact financial benefits that accrue to the U.S. LCA sector, it is necessary to explain in more detail the different aspects of NASA's aeronautics programs, (see figure 4.2 above). Aeronautical Research and Technology (ART) is the generic term for NASA's overall aeronautical activities. ART is subdivided into the Research and Technology Base Program (R&T Base) and the Aeronautical Focused Programs. R&T Base consists of generic research activities, including the following:

- ➢ Aerodynamics
- ➢ Structures
- ➢ Materials
- ➢ Human factors.

In the last six years NASA's overall budget allocation for R&T Base has been:

Table 4.2 R&T Base Budget (U.S. $ in million)

FY93	FY94	FY95	FY96	FY97	FY98	Total
436.5	448.3	366.3	354.7	404.2	418.3	2428.3

(Source: NASA budget, 1998)

NASA defined six strategic goals for its R&T Base program:

1. Develop high-payoff technologies for a new generation of environmentally compatible, economically superior U.S. subsonic civil aircraft and a safe, highly productive global air transportation system;
2. Prepare the technology base for an economically viable and environmentally friendly high-speed civil transport;
3. Establish the technology options for new capabilities in high-performance aircraft;
4. Develop and demonstrate technologies for hypersonic airbreathing flight;
5. Develop advanced concepts, physical understanding, and theoretical, experimental, and computational tools to enable advanced aerospace systems; and,
6. Develop, maintain, and operate critical national facilities for aeronautical research and for support of industry, FAA, DoD, and other NASA programs.

(Source: NASA Office of Aeronautics)

All these elements have an added objective of developing multidisciplinary methods that will contribute to U.S. industry's goal of reducing design cycle time by 50%, driven by the need to reduce product costs and capture increased market share.

According to the NASA Office of Aeronautics, the following technologies developed via the R&T Base Program can be found on specific U.S. LCA:

- Supercritical Wing for the B-757 and B-767;
- Winglets for the MD-11 and B-747-400;
- Acoustic nacelles for the MD-11, B-757, B-767, and B-747;
- Active turbine cooling for the JT9D engine and the B-747;
- Composite structures and advanced aluminium alloys for the B-757, B-767, B-747, and the MD-11;
- Advanced flightdeck displays for the B-757, B-767, B-747, and B-777.

(NASA Office of Aeronautics, FY 1996 Budget Report).

Despite these contributions to specific American LCA programs, the U.S. government does not consider R&T Base funding a subsidy to the U.S. LCA sector. But the Director of Aerospace and Science Policy at the American Institute of Aeronautics and Astronautics defines the goal of R&T Base as follows:

> The Research and Technology (R&T) Base is the essential element in opening new opportunities for future aeronautics advances that will contribute to improved performance, enhanced safety, reduced costs, higher reliability, and better operability of U.S. aviation systems. The seeds of tomorrow's competitiveness for the U.S. aviation lie in the NASA R&T base ... The next generation of systems technology programs that will follow the HSR and AST programs will come from the R&T base. It is the source of the expertise needed to solve the problems we can't foresee in the near term, (Testimony to the House of Representatives, February 10, 1994).

Aeronautics Focused Programs

The second major components of the ART are the Aeronautical Focused Programs, such as the High Speed Research Program (HSR) and the Advanced Subsonic Technology Program (AST). These programs investigate advanced aeronautics concepts in order that technological breakthroughs may be incorporated into current and future U.S. LCA programs.

The Advanced Subsonic Technology Program

The AST program is focused on technology transfer for a new subsonic U.S. civil jet transport that is aimed to regain technological leadership from Airbus Industrie. In recent years the AST budget has increased almost 20 fold.

Table 4.3   **NASA AST Budget (U.S. $ in million)**

	FY93	FY94	FY95	FY96	FY97	FY98	Total
AST	12.4	101.3	150.1	169.8	173.6	211.1	818.3

(Source: NASA Budget, 1998)

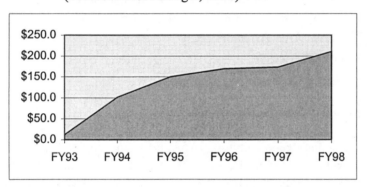

The funding increase for AST is a consequence of the changing priorities of NASA under Administrator Daniel Goldin. In keeping with the tenor of the Clinton Aviation Initiative, NASA's Focused Programs are now a key government strategic priority. As Wesley Harris, NASA's Associate Administrator explained to the House of Representatives:

> NASA's objective in the Advanced Subsonic Technology program is to provide U.S. industry with a competitive edge to recapture market share, maintain a strongly positive balance of trade, and increase U.S. jobs, (Testimony to the House of Representative, February 10, 1994).

This policy orientation was also clearly spelled out by the National Research Council in a major 1992 report on *Aeronautical Technologies for the 21st Century*. This report explains in detail the economic losses if the U.S. Government fails to help its large commercial aircraft industry in the intensifying competition with Airbus Industrie. It also specifies why NASA should be given the financial tools to develop the technologies needed to develop future subsonic and supersonic aircraft, (National Research Council, 1992). The AST program seeks to achieve technological breakthroughs in a number of areas, such as:

> Fly-by-Light/Power-by-Wire;
> Composites;
> Integrated Wing Design;
> Propulsion;
> Noise Reduction;
> Technology Integration and Environmental Impact;
> Environmental Research Aircraft and Remote Sensor Technology;
> Terminal Area Productivity;
> Shorthaul Aircraft;
> Ageing Aircraft.

The High Speed Research Program

In the 1990s the second major NASA Focused Program was the High Speed Research Program (HSR). Although terminated in 1999 because Boeing was unwilling to fund a demonstrator, the program has seen more than one billion dollars spent on developing technologies for a supersonic airframe. As we detail below the program included a contract to Boeing and McDonnell Douglas for US$440mn. The overall HSR budget profile is indicated in table 4.4 below.

Table 4.4 NASA HSR Budget (U.S. $mn)

	FY93	FY94	FY95	FY96	FY97	FY 98	Total
HSR	117.0	187.2	221.3	233.3	243.1	245.0	*1036.9*

(Source: NASA budget, 1998)

The 1990s HSR was not the first commercial supersonic program developed in the United States. In the 1960s, the U.S. Government had already funded a supersonic transport (SST) program, but this was cancelled in 1971, as it did not resolve issues over economic viability and environmental acceptability. More importantly, the program failed to deliver the required technology. The program was then resuscitated in 1987 and NASA was asked by the Office of Science and Technology to investigate the High-Speed Civil Transport (HSCT) technical feasibility, economic viability, and environmental compatibility, as well as to identify potentially high-payoff technology developments for HSCTs.

In 1989, McDonnell Douglas (MDC) and Boeing answered a NASA inquiry, and determined that a potential market existed for an HSCT. Their reports concluded that the stakes for supporting or abandoning high-speed advanced aircraft technologies were high, and could make a substantial difference in balance of trade benefits.

The market targeted by the HSR program was the Pacific Rim. The 1989 MDC and Boeing reports predicted that a projected quadrupling of traffic to and from the nations of this area, together with more moderate increases in demand for long range passenger transportation in other areas of the world, would create a need for some 500 next generation supersonic transports, worth an estimated $200 billion and likely to create 140,000 jobs.

In 1999 Boeing terminated the program because of changed economic circumstances in Asia. and because of the company's reluctance to self-fund a demonstrator. However, technology developed during the HSR will certainly be used in future Boeing aircraft projects. Also the proposed supersonic project can be easily resurrected.

Model of NASA's Supports to the U.S. LCA Sector

In the preceding analysis of NASA programs, "top down" expenditure figures were taken from NASA's budget and budget requests. However, these lack three key details necessary to assess the actual amount of expenditure that supports the U.S. LCA sector. These three factors are:

- ➢ Determination of Research and Program Management budget
- ➢ Distinction of civil from military R&T

> ➤ Discrimination of expenditure going only to the U.S. aerospace industry.

In this section we will indicate how we have estimated the relevant Research and Program Management budget for aeronautics and how we have determined the division between civil and military applications within NASA R&T. Further, we show how we calculated the value of expenditure going only to the aerospace industry.

Regarding Research and Program Management (RPM) it needs to be recognized that NASA does not include the intramural costs of running the Aeronautical Research and Technology programs within its definition of the aeronautics budget. Yet during the peak years of the HSR about 650 government scientists and engineers were working on the program with Boeing and McDonnell Douglas.

NASA's budget proposals are included in the overall budget proposal of the U.S. government. However, in order to gain accurate figures for the actual financial outlays going to NASA data must only be taken only after proposals have passed through each house of Congress and the Congressional Conference Committee. This is illustrated in chapter four.

In order to obtain a rounded view of NASA expenditure, it is necessary to utilize figures from the NASA Office of Aeronautics for areas such as Research and Program Management, (RPM). Similarly, data from the General Accounting Office (GAO) can serve as another crosscheck. Nevertheless, two problems remain which will impact on the degree to which one can monitor U.S. compliance with current and future disciplines on LCA subsidy.

1. Dollar values for NASA R&D contracts are provided in statements for the preceding fiscal year. Thus values for 1992 are given in Federal accounts for 1993. However, when one moves forward in time to a later fiscal year the same contracts often show lower values. The U.S. GAO recently highlighted these difficulties in a report:

... Because of the government's systems, record keeping, documentation, and control deficiencies, amounts reported in the consolidated financial statements and related notes do not provide a reliable source of information for decision-making by the government or the public. These deficiencies also diminish the reliability of any information contained in any other financial management information – including budget information and information used to manage the government day-to-day – which is taken from the same data sources as the consolidated financial statements, (U.S. GAO, Results of Fiscal Year 1997 Audit, April 1998).

2. Federal agencies, such as NASA, call their budget proposals "estimates". With regard to program area budgets NASA requests may not correspond to what Congress actually approves. More problematically, the actual amount NASA spends may be entirely different. If appropriated funds are not spent in their entirety, these may be carried over to support future resource gaps in later years. Remarkably, at the beginning of FY 1996, NASA had $3.6 bn in "carryover" balances.

We mention these accounting issues here to warn the reader that the crosschecking of the accuracy of data on federal funding may well show inconsistencies. Because of this the analysis here is based wherever possible on actual expenditure, (for more detail see the discussion of our methodology in Appendix A). Figures for actual expenditure can also be found in the annual report of the Aerospace Industries Association (AIA), (Aerospace Industry Association, 1998/99, p. 105).

NASA's Overall Budget Structure

NASA's Aeronautics Program budget is a subdivision of the total NASA budget, which includes funding for space, operations, research and program management and construction of facilities. Until 1994 NASA's internal classification of its budget was as follows:

➤ Research & Development
➤ Space Flight & Data Communications
➤ Research & Program Management
➤ Construction of Facilities.

After 1995 NASA changed its budget classification to the four subdivisions listed below:

- ➢ Science, Aeronautics & Technology
- ➢ Human Space Flight
- ➢ Mission Support
- ➢ Other.

Fortunately, in this case, the changes to the budget classification are largely cosmetic. Mission support consists largely of what was formally the sum of Research & Program Management and Construction of Facilities. While the relevant Aeronautical Research and Technology budget titles were previously listed under Research and Development but are now found under Science, Aeronautics & Technology.

A more complex issue arises in assessing the overall outlay for aeronautics. Broadly speaking, the overall structure of the budget can be divided into aeronautics and space, with by far the largest sums spent on space related activities. But, as we have argued above, the proportion attributed to aeronautics must include a percentage of the funding given to Research & Program Management (R&PM) and Construction of Facilities. In previous reports on U.S. federal funding for the American LCA sector, the issue of Research and Program Management expenditure has been controversial. As far as we can ascertain, the official U.S. view of this expenditure is that it is for internal government functions. But this is specious. As we illustrated above with the HSR program, the personnel dedicated to operating aeronautics R&T programs are integral to NASA's aeronautics mission to assist U.S. aerospace companies. The salaries of such personnel can thus be legitimately included in the NASA expenditure for aeronautics. To reinforce the point, it should be noted that the U.S. Aerospace Industries Association (AIA) includes a percentage of the R&PM budget in its calculation of the overall NASA outlay for aeronautics, (AIA, 1998/99, p.105).

Aeronautics Research & Program Management Expenditure

In order to establish figures for aeronautics R&PM, we utilized NASA data on full-time equivalent workyears (FTE's) for the fiscal years 1996, 1997 and 1998. We approached the issue from this angle because neither Congressional nor NASA budget sources provide actual cost estimates for the subdivisions of the Research & Program Management budget. We also referenced this against data for 1985 in order to ascertain whether the recent trends were compatible with earlier NASA practice, which they are. However, it should be noted that this data included figures for Safety, Reliability & Quality Assurance, Construction of Facilities, Program Management and Center Operations. On the basis of the available information we concluded that the aeronautics related FTEs amounted to 21.6% of the total equivalent workyears. This figure was then also applied to the total R&PM expenditure, since it mainly consists of personnel and related costs.

Table 4.5 Distribution of NASA Full-time Equivalent Workyears

	FY 1996	FY 1997	FY 1998	Average %
Aeronautical R&T	3284	3425	3440	
Space R&T, Dev	12621	12461	11750	
Subtotal	**15905**	**15886**	**15190**	
Safety, Reliability, Quality Assurance	124	105	105	
Construction of Facilities	224	176	172	
Program Management H.Q.	554	55	55	
Centre Management & Operations	3978	3714	3523	
Other controlled FTEs	652	565	514	
Total FTEs	**20938**	**20501**	**19559**	**100%**
(of which) Total Aeronautics	4321	4422	4427	21.6%
(of which) Total Space	16617	16079	15132	78.4%

The detailed figures for NASA's internal allocation of FTEs to program areas are given for FY 1996, FY 1997 and FY 1998 above in table 4.5. On this basis we estimated that 21.6% of the total NASA R&PM budget was allocated to aeronautics. The results of the analysis for calculating Aeronautics Research and Program Management funding are given below in table 4.6. As the reader will see the expenditure amounts to more than US$300mn per year for the specified period. As far as the aeronautics-relevant Construction of Facilities expenditure is concerned, details were taken from the extensive list of sources provided in Appendix B. The most significant source of information here was the NASA Fiscal Year Authorization Acts. These details allowed for a categorization by sector, such that it was possible to attribute the appropriate funding quantities to space or aircraft on the basis of the information provided.

Table 4.6 Estimated Expenditure: NASA Aeronautics Research and Program Management (Current US$mn)

Year	Research and Program Management	Total
1992	303	303
1993	309	309
1994	300	300
1995	309	309
1996	314	314
1997	345	345

Civil/Military/Dual-Use Balance of NASA Aeronautics Programs

As with the R&PM budget, the breakdown of the subdivisions of the Aeronautical R&T generic program created a significant methodological issue for the authors of this study. NASA's own classification of discrete research fields does not specify the relevance of a given area of research for military or civil applications. In consequence, we decided to convene a panel of experts from companies, universities and research establishments to give a judgement on the percentage of different NASA program areas which could be construed as either civil, military or dual-use. Our purpose was to exclude any research activity that could not have a potential U.S. LCA application. This qualitative exercise, which involved several weeks of workshops at the European Association of Aerospace Industries (AECMA) HQ, complemented the quantitative data taken from Congressional sources. (Details of panel members are available in Appendix B).

The percentages presented below in table 4.8 were applied to both Aeronautical Research and Technology Expenditure and Research and Program Management funding. However, they were not used to estimate Aeronautical Construction of Facilities expenditure, since the categorization by Civil/Military/Dual-Use was already possible on the basis of the referenced documents found in Appendix B. Table 4.7 shows the breakdown of NASA aeronautical expenditure by sector after the percentages given in table 4.8 are applied.

Table 4.7 Estimated Expenditure for NASA Aeronautical R&T by Sector (Current US$mn)

Year	Purely Civil	Purely Military	Dual-use	Total
1992	429	64	150	642
1993	488	73	168	728
1994	609	91	209	909
1995	493	74	169	736
1996	512	76	176	764
1997	562	84	193	838

Table 4.8 Expert Analysis of Civil/Military/Dual-Use Percentage of NASA Programs

PROPOSED CATEGORISATION	PURELY CIVIL	PURELY MILITARY	DUAL USE	1996-98 Average %
Research & Technology Base [47.7 % of Total Aeronautics R&T]				100%
Information Technology	0 %	0 %	100 %	18%
Airframe Systems	65 %	35 %	0 %	33%
Propulsion Systems	60 %	20 %	20 %	19%
Flight Research	30 %	30 %	40 %	18%
Aviation Operations Systems	0 %	0 %	100 %	4%
Rotorcraft	0 %	0 %	100 %	8%
	38.2 %	20.8 %	41 %	
Aeronautical Focused Programs [52.3 % of Total Aeronautics R&T]				100%
High-performance Computing And Communications	0 %	0 %	100 %	7%
High-Speed Research	100 %	0 %	0 %	52%
Advanced Subsonic Technology	100 %	0 %	0 %	41%
	93 %	0 %	7 %	
[Using 47.7 % and 52.3 % Weighting Factors, Respectively]				
Total	67 %	9.8 %	23.2 %	

(Source: Aerospace Strategy Research Centre/AECMA panel members).

Destination of NASA R&D Contracts

By using figures from the NASA annual budget report, we were able to further analyze NASA aeronautics expenditure in relation to the destination of R&D contracts. These contracts are the chief means through which NASA provides benefits to the U.S. large commercial aircraft sector, as the research results are given to the companies which are also paid to undertake the research Our analysis reveals that 65% of all aeronautics R&D contracts go to U.S. aerospace companies and 35% are awarded to U.S. companies that are not primarily in the aerospace sector, as well as U.S. universities and other government institutions.

The entirety of the assumptions enumerated above characterizes our methodology for assessing NASA aeronautics expenditure. This is illustrated in the schematic figures 4.4 and 4.5 below.

On the basis of the assumptions illustrated in figures 4.4 and 4.5 we are able to present "top down" data to show the funding quantities within the overall NASA budget and the subdivisions of aeronautics. These can be found in *Detailed Expenditure for NASA Aeronautics* (tables 4.9 to 4.13 below). Although these data are "top down" a number of methodological devices have been used to track the flow of expenditure into the U.S. LCA sector. Civil/military/dual-use distinctions were established through a detailed investigation of NASA's internal operations. The aim was to exclude military-only expenditure. Some minor differences between maximum and minimum figures illustrate areas where the available information could not support cast iron designations of program character and content. With regard to assessing benefits to the U.S. LCA sector, our analysis of NASA contracts allowed a categorization of subcontracted work by NASA to the U.S. LCA industry by the value of contracts awarded.

Figure 4.3 Schematic of Methodology to Estimate NASA Aeronautics Expenditure

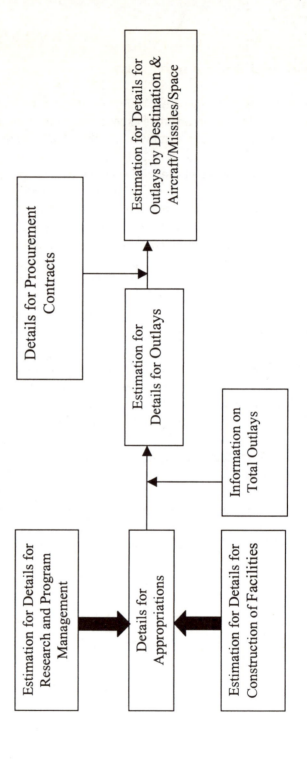

Figure 4.4 Schematic of Methodology to Estimate NASA Expenditure for Aeronautics: How Much of $\bigotimes$ is LCA Relevant?

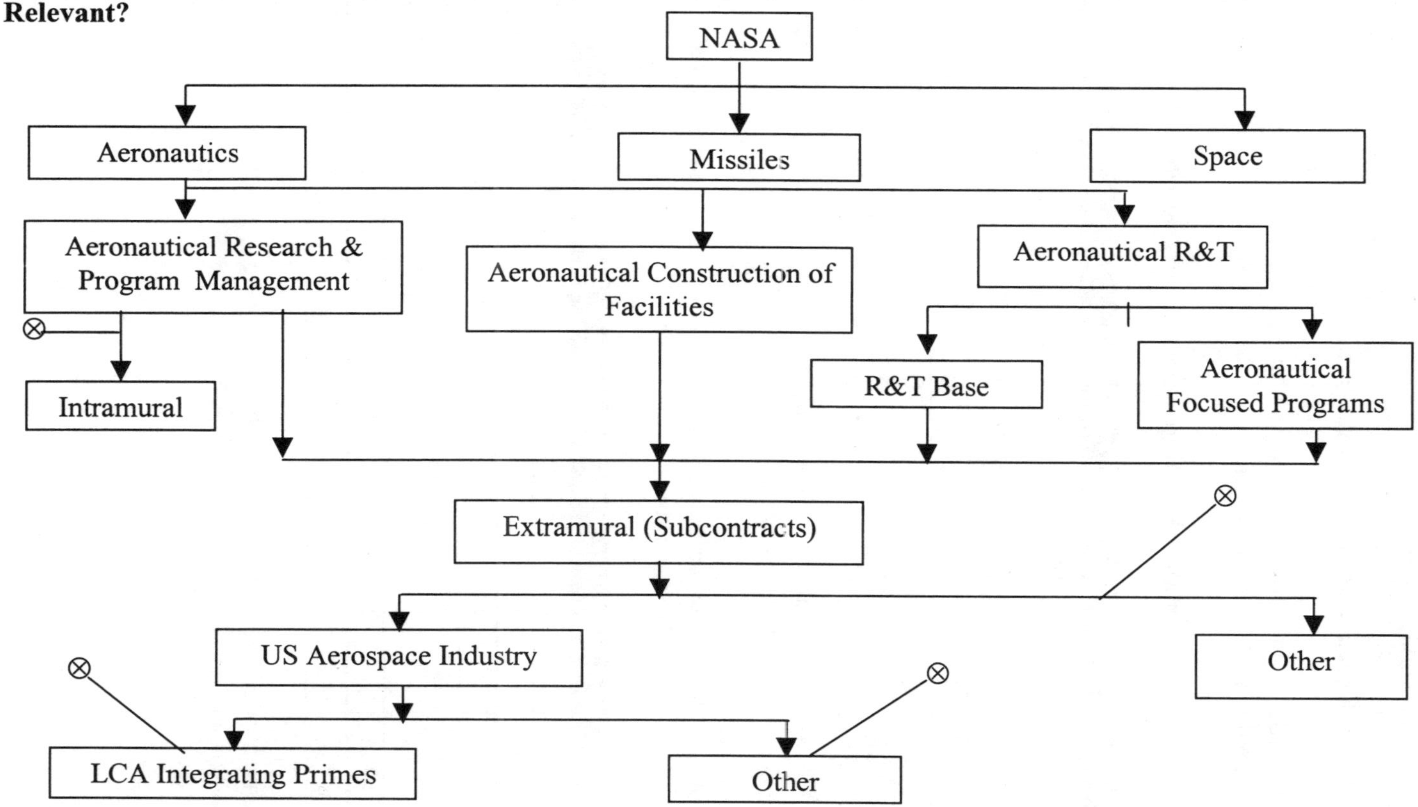

Detailed Expenditure for NASA Aeronautics

The reader should note that minimum/maximum figures have only been included where the figures differ.

Table 4.9   Total NASA Expenditure (Current US$mn)

Year	Aeronautics	Missiles	Space	Total
1992	1125	0	12835	13960
1993	1212	0	13092	14304
1994	1330	0	12363	13693
1995	1153	0	12593	13746
1996	1187	0	12694	13881
1997	1302	0	13055	14357

(Due to rounding and the selected methodology (see Appendix A) details do not always add up to totals).

Table 4.10   Estimated Expenditure: NASA Aeronautical R&T by Destination (Current US$mn)

Year	Intramural	To US Industry	To Other	Total
1991	0	337	182	519
1992	0	418	225	642
1993	0	473	255	728
1994	0	591	318	909
1995	0	478	257	736
1996	0	497	267	764
1997	0	545	293	838

(Due to rounding and selected methodology (see appendix A) details do not always add up to totals).

Table 4.11   Estimated Expenditure: Subcontracted work from NASA Aeronautics (Current US$mn)

Year	Purely Civil		Purely Military		Dual Use		Total	
	Min'm	Max'm	Min'm	Max'm	Min'm	Max'm	Min'm	Max'm
1992	499	528	79	108	215	215	822	822
1993	557	593	87	123	223	223	903	903
1994	671	690	101	120	239	239	1030	1030
1995	555	567	83	95	193	193	844	844
1996	575	590	86	101	197	197	873	873
1997	631	647	94	110	216	216	957	957

(Due to rounding and selected methodology (see Appendix A) details do not always add up to totals).

Table 4.12   Estimated Expenditure: NASA Subcontracted Work to U.S. Aerospace Industry (Current US$mn)

Year	Research and Program Management	Construction of Facilities	Aeronautical R&T	Total
1992	59	58	418	534
1993	60	53	473	587
1994	57	20	591	669
1995	60	10	478	548
1996	61	10	497	567
1997	67	11	545	622

(Due to rounding and selected methodology (see Appendix A) details do not always add up to totals).

Table 4.13   Estimated Expenditure: NASA Subcontracted Work to U.S. Aerospace Industry by Sector (Current US$mn)

	Purely Civil		Purely Military		Dual Use		Total	
Year	Min'm	Max'm	Min'm	Max'm	Min'm	Max'm	Min'm	Max'm
1992	324	343	47	70	140	140	534	534
1993	362	385	53	80	145	145	587	587
1994	436	448	65	78	155	155	669	669
1995	361	369	54	62	126	126	548	548
1996	374	383	56	65	128	128	567	567
1997	410	420	61	72	141	141	622	622

(Due to rounding and selected methodology (see Appendix A) details do not always add up to totals).

Summary of Detailed Estimations for NASA Aeronautics Expenditure

This section provides a summary of NASA aeronautics expenditure given in more detail above in tables 4.7 to 4.13.

Table 4.14   NASA Total Aeronautical Expenditure 1992-1997, (Current U.S. $mn)

Year	1992	1993	1994	1995	1996	1997	*Total*
U.S.$mn	1125	1212	1330	1153	1187	1302	*7309*

Table 4.15   NASA Subcontracted work to the U.S Aerospace Industry (Current U.S. $mn)

Year	1992	1993	1994	1995	1996	1997	*Total*
U.S. $mn	534	587	669	548	567	622	*3527*

Table 4.16   Value of Civil and Dual-Use Subcontracted Work to the U.S. Aerospace Industry (Current U.S. $mn)

Year	1992	1993	1994	1995	1996	1997	Total
U.S. $mn	464	507	591	487	502	551	3102

Figure 4.5   Value and Flow of NASA Funding into U.S. LCA Sector

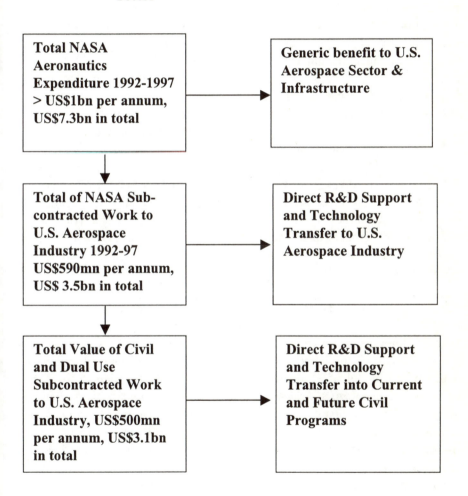

Summary

The expenditure totals listed above are derived from an exhaustive search of U.S. data bases on government budgets. The data is thus typically of the "top down" variety. However, by using a number of methodological tools we have been able to refine the analysis by pinpointing precise destinations for funding. In previous studies of U.S. Government funding of the U.S. LCA sector the inability to track funding precisely has been a major weakness. Thus we have distinguished military, civil and dual-use applications as well as intramural expenditure versus funding on external R&D contracts. Through an analysis of NASA contracts we were also able to ascertain the percentage of sub-contracted work (externally awarded contracts) going to U.S. aerospace companies. We believe that these figures are more accurate than the program specific budgets that are the normal basis for studies on government aerospace expenditure.

On the basis of our analysis we have identified very large sums of federal spending on aeronautics in the United States. In the six years from 1992-1997, NASA's total expenditure on aeronautics was more than $1billion a year, totalling $7.3 billion overall, (table 4.14). Of this amount more than $500million per year, $3.5 billion in total, was spent on externally subcontracted work to the U.S. aerospace industry. Total civil and dual-use contracts to the aerospace industry were worth $4.1 billion, (table 4.17). A flow chart depiction of the value of industry subcontracts is given above in figure 4.5.

U.S. LCA Sector R&D Contracts: a "Bottom Up" Analysis

In the following section we will provide an analysis of NASA contracts to the U.S. LCA sector based on a picture established through building a "bottom up" model of the value of NASA contracts awarded to the U.S. LCA sector. In *Detailed Expenditure for NASA Aeronautics* we indicated overall NASA aeronautics expenditure and the value of work sub-contracted to the U.S. aerospace industry using "top down" budgetary and contract data. However, these contracts are sometimes awarded to companies not directly involved in LCA manufacture. As a result in this section we have sought more detail by investigating the NASA database on contracts which were awarded only to Boeing and McDonnell Douglas.

Our aim is thus to provide a "bottom up" picture of NASA's financial benefits to the U.S. LCA sector by isolating individual contracts given to the manufacturers. However, as the NASA database is purged every year we were only able to select contracts that were still active in 1996/1997, although some of these go back as far as 1989. Contract details are given in Appendix D (table 4.9).

The list of U.S. LCA relevant NASA contracts has been categorized by Purely Civil/Purely Military/Dual-Use, and by a keyword describing whether the contract is believed to belong to the Research and Technology Base or the Aeronautics Focused Program. We believe that we have identified most, if not all, of the Aeronautics Focused Program contracts, but some of the R&T Base contracts have not been clearly identified, particularly those with a dual-use potential. Thus table 1 and 2 in Appendix D exclude many R&T Base contracts. In consequence our estimate of financial benefit to the U.S. LCA sector is *only a minimum figure*.

Because contracts extend over several years the totals cannot be allocated to a single year. Accordingly, we have distributed expenditure over the duration of the contracts by a left skewed normal distribution curve, (see table 5 in Appendix D). By using this method we have provided average expenditure figures for the years 1996/1997. Having calculated average contract data for 1996/97 the contracts were categorized as either civil/military/dual-use in order to exclude military-only R&T. Calculating the values for earlier years is impossible because of lack of contract data.

Because the recipients of contracts are known we were able to compare the "bottom up" data with our "top down" data on NASA expenditure. In our analysis of contracts we found that 65% were awarded to the U.S. aerospace industry, but of these a small proportion went to companies other than Boeing and McDonnell Douglas. Our "bottom up" picture is thus compatible with the "top down" view, as according to our estimation roughly 50% of AFP expenditure went directly to the two U.S. LCA manufacturers in 1996/97. For the Aeronautical Focused Programs, the results are shown in table 4.17 below.

Table 4.17   Average Annual Value of Contracts to U.S. LCA Prime from the NASA Aeronautical Focused Programs for 1996/97 (Current US$mns)

	LCA Primes	Other	Total
Purely Civil	193	0	193
Dual-Use	11	0	11
Total	204	0	204

On the basis of our analysis we believe that an annual average of $204mn has been transferred to the U.S. LCA manufacturers in both 1996 and 1997 via AFP programs

The same exercise used in the Aeronautics Focused Programs was also applied to NASA R&T Base Contracts. The totals are given in table 4.18 below. We estimate the R&T Base Program's financial benefit to the U.S. LCA manufacturers for each of the years 1996/1997 to have been U.S.$186mn.

Table 4.18   Average Annual Value of Contracts to U.S. LCA Primes from the NASA R&T Base Program for 1996/97 (Current US$mns)

	LCA Primes	Other	Total
Purely Civil	150	0	150
Dual-Use	36	0	36
Total	186	0	186

Adding the R&T Base and the Focused Program contracts together the total minimum financial contribution from NASA R&D contracts to the U.S. LCA primes is thus calculated to be U.S.$390mn for both 1996 and 1997.

Table 4.19 **Average Annual Value of R&T Base and Focused Programs Contracts to U.S. LCA Primes for 1996/97 (Current US$mns)**

	LCA Primes	Other	*Total*
Purely Civil	343	0	*343*
Dual-Use	47	0	*47*
Total	*390*	*0*	*390*

The values given above in tables 4.17 to 4.19 exclude any contribution from Research and Program Management or Construction of Facilities Expenditure, (see NASA's Overall Budget Structure above). By adding the relevant sums from those categories we increase the total average value of contracts to the LCA prime companies for 1996/97 to US$mn433, (for data on research and program management see table 4.6).

The Nature of Subsidy: Applications of NASA R&D

In the preceding sections an analysis has been presented of the scale of NASA funding of the U.S. LCA sector. This now needs to be linked to the qualitative question of how the R&D funding actually benefits the U.S. large commercial aircraft sector. At the beginning of this chapter we provided an overview of NASA's history and mission and recent federal policy for aeronautics. Our aim was to link NASA's policy orientation to the broader aspects of U.S. aeronautics R&D policy. We contend that NASA's major goal in its aeronautics programs is to provide the U.S. LCA manufacturing sector with long and medium range R&T in order to strengthen the science and technology base of U.S. companies working in the LCA sector. In essence, the public money spent on R&D is to assist companies to attain the knowledge competencies necessary for U.S. competitiveness. Regarding the question of subsidy, it is important to realize that technologies developed at government expense in NASA are repeatedly incorporated into U.S. LCA programs.

Our conclusion regarding the role of NASA and Federal funding is echoed in a number of U.S. official reports which we have scrutinized. A key one is The *NASA/DoD Aerospace Knowledge Diffusion Research Project* which notes that:

> Collectively, the results of the R&D performed and sponsored by federal agencies constitute the largest contribution to the U.S. aeronautics knowledge base... NASA conducts and sponsors research across a broad spectrum of aeronautics and space technology, including aerodynamics, acoustics, aeroelasticity, avionics, computational fluid dynamics (CFD), controls, materials and structures, propulsion and propulsion integration, and flight management. With respect to aeronautics, U.S. public policy is directed at maintaining the competitive position, now and in the future, of the U.S. aerospace industry, (Barclay & Pinelli, 1997, p. 917).

Table 4.20   U.S. Commercial Aircraft Incorporating NACA/NASA R&T

Aircraft	NACA/NASA R&T Incorporated
Ford Tri-Motor	Wing contour and wing cowlings developed by NACA to improve airflow characteristics
Douglas DC-3	Low-drag engine cowling design, developed by NACA
Lockheed Constellation	NACA's aerodynamic drag reduction experimental research results and the low-drag engine cowling
Boeing 747	NASA research in high-bypass jet engines, low drag nacelles, swept-wing, airfoils, noise reduction, transonic aerodynamics, and structural research
McDonnell-Douglas MD-11	Winglet design, supercritical airfoils, digital electronic controls, numerous engine design improvements, high-lift systems, transonic aerodynamics and structural concepts
Boeing 777	Digital flight controls, glass cockpit, quiet engine nacelles, aerodynamic design codes, flight management systems, graphite-epoxy structures and transonic supercritical airfoils

(Adapted from the NASA/DoD Aerospace Knowledge Diffusion Project, Pinelli et.al, 1997, p. 37).

In order to reinforce the points made in the quotation above we show above in table 4.20 the different NACA/NASA technologies incorporated into U.S. made aircraft. It is interesting to note that two of Boeing's leading products, the B-747 and the B-777, have benefited substantially from NASA technologies. If we further consider that the B-707 received substantial government funding via the KC-135 program, then we can see that Boeing's commercial division has benefited substantially from partnerships with government agencies.

NASA's U.S. LCA Subsidy and R&D Contracts

Many of the technologies incorporated into U.S. large commercial aircraft come from work done on NASA contracts by engineers from the U.S. LCA sector. It must be realized that this is not the normal form of a commercial contract. Normally a contract means that goods or services exchange hands for a sum of money. But in the case of NASA R&D contracts, the work produced as a service leads to the generation of knowledge and technology which is then given to the companies which were paid to do the research. In our view this clearly conveys a financial benefit.

An excellent example of this benefit can be found in a contract awarded in September 1995. NASA awarded contract NAS 1 20546 to McDonnell Douglas to develop an 'Advanced Composite Technology Wing', the contract was worth $121,861,556 and was funded jointly from the Advanced Subsonic Transport and High Speed Research Programs. The contract will run until 2001, (after the Boeing/McDonnell Douglas Merger in 1997, the contract was absorbed by Boeing). Financial payments for the contract are indicated in table 4.21 below.

Table 4.21   Annual Funding of NASA Composite Wing Contract (U.S.$000s)

FY	1995	1996	1997	1998	1999	*Total*
U.S. $000s	508	13,507	21,204	22,600	16,558	*74,337*

(Source: U.S. General Services Administration).

As we can see, Boeing/MDC had received more than $74million for work on this contract by 1999. Of more interest is the fact that the contract has

81

already led to the fabrication of a 40-foot wing at the Boeing/MDC Phantom Works in Long Beach California. In September 1999 the wing was taken to NASA Langley for testing, which included a maintenance analysis by American Airlines mechanics, (NewsEdge Corporation, 10/9/99, p. 2).

As we saw in chapter one composite materials are the key to lighter aircraft structures, however, their use has been held back by high manufacturing costs. The wing produced on the NASA contract detailed here already shows a 25-30% weight saving, with a potential direct operating cost (DOC) saving of between 4% and 10%. In terms of applications the new wing could be fitted to current aircraft or to new programs under development by Boeing, such as a blended wing, very large civil transport.

Regarding the overall picture it must be realized that this technology contract is not unique. By looking in detail at other NASA contracts we have ascertained that almost all the large NASA R&D contracts undertaken since 1989 have been awarded to U.S. LCA primes. For example:

> 1994 High Speed Research Airframe: Boeing U.S. $ 440mn
> 1995 ACT Wing Technology: McDonnell DouglasU.S. $122mn
> 1997 Advanced Composite Technology: Boeing U.S. $130mn.

Contracts awarded to companies that are main suppliers to the primes are also of major benefit to the U.S. LCA sector. Systems and components which are not produced by the integrating primes are a crucial part of a modern LCA and now represent more than 50% of the overall value of an aircraft. Many contracts of substantial value are awarded to companies in the U.S. LCA supply chain, such as Honeywell, Sundstrand and Dow Chemicals. Others are awarded to consortia that often include Boeing. Thus it is not implausible to suggest that virtually all the funds allocated to NASA R&T, which are of civil sector relevance and are spent on contracts given to U.S. companies, represent a financial contribution to the U.S. LCA sector.

Distribution of NASA R&T

It is often claimed that NASA research results are put in the public domain and are not of benefit to specific U.S. companies. But the results of R&T Base work, including research not undertaken by engineers from the U.S. LCA primes, are distributed to the aircraft industry through NASA technical reports, NASA sponsored workshops and symposia. The work is conducted in unique government funded facilities, which are not available in private industry or universities, (National Academy of Sciences, 1995, p. 17.). A good example of the special nature of NASA's resources is the Numerical Aeronautical Simulation (NAS) Facility at NASA Ames, which gives U.S. engineers access to one of the world's most powerful supercomputers, (COTA, 1991, p. 70.) With respect to the NAS facility, it should be noted that NASA guards its use very closely. NAS is open only to U.S. companies. NASA Ames also monitors university use very closely in order to prevent foreign students gaining access to U.S. technology, (COTA, 1991, p. 74). More broadly, the dissemination of important research results can be controlled by NASA through the For Early Domestic Distribution rule (FEDD). By using the provision of FEDD, NASA can withhold external distribution of research results by up to three years. Technical reports stamped FEDD are presented through briefings and presentations exclusively to U.S. firms. As well as the FEDD procedure, another legal means to ensure the domestic retention of intellectual property is the Limited Exclusive Rights Data (LERD) restriction, which was developed specially for the HSR and AST programs. LERD provides the legal basis for the exchange of information on these programs, while ensuring that the technology is 'protected from foreign interest', (Intellectual Property Protocol of Limited Exclusive Rights and Project Sensitive Data, NASA Glenn Research Centre, Procedure 22, April 1999).

Providing knowledge and technology to U.S. companies is straightforward. On the Aeronautical Focused Programs distribution of R&T is very simple. As the *DoD/NASA Aerospace Knowledge Diffusion Report* notes, 'Dissemination of Focused Program research results is limited to the "circle of friends" who participated, (Pinelli et.al. 1997, p. 45).

Conclusions

NASA subsidizes the U.S. LCA sector in a number of ways. Generically, all of NASA's aeronautics expenditure is geared to securing the competitive position of the U.S. aircraft industry, particularly its LCA sector. NASA's aeronautics funding is aimed to provide industry support on a national mission basis and to secure a competitive edge in the global market place. This is not, in our view, contestable. Public documents, statements of NASA officials and the *NASA/DoD Aerospace Knowledge Diffusion* study all make this aspect of NASA's mission abundantly clear.

The LCA focused subsidy provided by NASA comes from R&D contracts that are undertaken in partnership with the U.S. aerospace industry, particularly its LCA sector. From these contracts, industry scientists and engineers take with them technology that is transferred onto U.S. LCA programs. In short, public funding transfers knowledge from NASA to the U.S. private sector. In the U.S. this is widely contested, but NASA's own documentation illustrates its contribution to U.S. LCA programs. As NASA Langley staff report:

> In May 1996, the first Boeing 777 stopped by Langley Research Centre as a salute to NASA's involvement in its creation. Several Langley innovations were instrumental in the development of the aircraft, *(NASA, Spin-Off 97, p.54)*.

The B-777, cited above, is an archetypal example of the competitive advantage which comes from the LCA public/private partnership where private companies derive benefit from public funding of R&D. The NASA/DoD Knowledge Diffusion Study notes:

> Some of the advances, including digital flight controls, the glass cockpit, quiet engine nacelles, flight management systems, graphite-epoxy structures, and transonic supercritical airfoils, are outgrowths of the publicly funded R&T conducted by the National Aeronautics and Space Administration *(NASA)*, (Kay, Pinelli and Barclay, 1997, p. 86).

84

The same authors continue:

> In many ways, the Boeing 777 is also a public policy success...The growth of the U.S. aircraft industry, especially the LCA sector, has been the result of a strong partnership between government, industry and academia, (1997, p. 86).

The clear fact of public/private partnership that characterizes U.S. LCA is routinely denied in U.S. policy-making circles. As a Department of Commerce official noted in 1997, 'The simple fact is that the U.S. Government does not provide subsidies for the development and production of civil aircraft – PERIOD. We just don't do it', (Interview of Ellis Mottur with the American Institute of Astronautics and Aeronautics, March, 1997).

In many pronouncements U.S. trade officials stress the independent and free market status of the U.S. LCA sector. But regarding this fundamental point concerning the question of subsidy let us leave the last word to a senior Boeing executive. In 1996, in a speech at NASA Langley, Boeing's vice president of engineering, Robert Spitze, emphasized the public/private partnership that marks his industry:

> The LaRC [Langley Research Centre] is the aeronautical birthplace of working together...Your relationships with other government research centers and academia have been quietly showing the way. The relevance and correctness of your working together is evidenced by aviation's incredible growth. Your working together success is evident in every U.S.-made aircraft, (quoted, Golich and Pinelli, 1997, p. 36).

No clearer statement about the role of publicly funded R&D in U.S. LCA is needed to sum up our analysis.

5 Department of Defense Subsidy of the U.S. Large Commercial Aircraft Sector

There has been very significant 'fallout' in civilian sector industries from this large DoD investment. It has resulted in the generation and/or rapid stimulation of major commercial industries (jet aircraft, computers, communication satellites etc)....stressing R&D has paid dividends not only for the DoD but also the US economy in general, (Gansler, 1990, p. 308).

The single greatest means by which U.S. government policy has affected the competitiveness of the commercial aircraft industry is in the procurement of military aircraft and funding of the related R&D. Of several ways in which the military side of the industry has advanced the commercial side, technology synergies are in the top rank of importance, (COTA, 1991, p.30).

Department of Defense R&D Policy

The generic term for NASA's research efforts is Research and Technology (R&T). However, in the Department of Defense matters are more complex. The DoD, as a customer of the U.S. aerospace industry, is heavily involved in the testing and evaluation of components, systems and platforms. Hence the DoD's generic term for its R&D activities is Research, Development, Testing and Evaluation (RTD&E). Accordingly, this terminology will be used in this chapter on the DoD. It should be noted that historically DoD's involvement in development and testing has heightened the level of benefits accruing to the U.S. LCA sector from defence/civil synergy, because systems and platforms developed at DoD expense have been spun off to the civil side. As leading industrial economists Mowery and Rosenberg note '... the history of technical development in commercial aircraft consists largely of the utilization for commercial purposes of technical knowledge developed for military programs at government expense', (1982, p.140).

Table 5.1   R&D Spending by Agency (U.S. $ mn)

Department	FY95	FY96	FY97
Defense	35,350	35,428	35,523
Health and Human Services	11,519	12,118	12,621
NASA (aeronautics & space)	9,390	9,334	9,359
Energy	6,481	6,689	6,269
National Science Foundation	2,431	2,430	2,516
Agriculture	1,542	1,479	1,499
Commerce	1,164	1,086	1,260
Interior	668	622	582
Transportation	667	622	679
EPA	554	508	585
Other	1,315	1,134	1,786
Total	*71,081*	*71,450*	*72,679*

(Source: U.S. Congress Authorization Budget).

The Department of Defense (DoD) implements and finances several RTD&E programs, executed either at a centralized level, through the efforts of the Defense Advanced Research Projects Agency (DARPA), or at the Services (Air Force, Army, Navy) level.

The scale of DoD funding for R&D relative to other agencies is illustrated above in table 5.1. The table displays the amount of money allocated by the U.S. Government by Agencies. These numbers are not specific to aviation R&D programs. They represent the entire Federal R&D budget. But what cannot escape notice is the huge annual budget of approximately $35bn for DoD RTD&E. Of this roughly $7bn is allocated to aircraft RTD&E, (1996 Aeronautics and Space Report of the President).

Of the many programs undertaken by the DoD, some have military-only applications, but other have a dual-use spin-off potential which allows the development of technologies that have easily transitioned to commercial applications, such as manufacturing technology, avionics and airframe development programs. In our analysis here great care has been taken to ensure that military only technology has not been included in our assessment of LCA relevant R&D expenditure. However, unlike our analysis of NASA, we have not been able to build a "bottom up" picture of U.S. LCA supports from DoD contracts. This is because some contracts

87

concern black, top- secret programs, but also because of the sheer numbers; some 70,000 contracts would have had to be scrutinized. Instead we have calculated the spin-off potential of DoD RTD&E and we provide case by case examples of commercial benefit from defence programs. An excellent example of this transition is given later in this chapter in the analysis of the role played by Boeing in the fabrication of the composite wing of the B-2 stealth bomber. Several other examples will also be cited in the following analysis.

Historical Dimensions of DoD R&D

For most of this century, the role of military research and development in fuelling technological change and, through technology transfer, creating competitive advantage in commercial as well as military markets has been profoundly important. Historically, in the U.S. military R&D has played an indispensable part in driving military capabilities. But in addition, through technology transfer to the commercial sector, DoD R&D has enhanced industrial productivity and the global competitive edge of the U.S. LCA sector. In mainstream economic analyses the role of military funded technology transfer has been underestimated and the pervasive impact of this process has not been fully appreciated. As David Noble argues:

> I would like to suggest that this conventional view of the role of the military in technological development is problematic on both counts. First, because the military role has not been the "externality" that it appears to be when viewed through the lens of the neoclassical economist. Rather it has been central to industrial development in the United States since the dawn of the industrial revolution.... Second, the influence of the military on technologies is not temporary, something removed when the technologies enter the civilian economy. The influence spills over in the specific shape of the technologies themselves and in the way they are put together and used, (Noble, 1987, p. 330).

In the aerospace industry synergies between military R&D and the technical requirements of the civil sector have been numerous and effectively exploited to sharpen U.S. industrial competitive edge in global markets.

Prior to World War I, the military budgets of governments funded the development of wireless, aviation and guidance and control systems. The Army and Navy provided support for the development of automatically piloted aerial torpedo prototypes and flying bombs. The driving force of military expenditure behind technological advance continued apace after 1918. In particular 'military-funded construction projects culminating in systems for the mass production of various military weapons also effected momentous technological changes' (Hughes, 1994, p. 426). The intense demand pressures on technology during World War II proved a massive stimulus to post-1945 technological advance.

In particular: 'wartime projects initiated three major technological systems that spread through the world in subsequent decades', (Hughes, 1994, p. 427.) These three critical technological systems included computers, nuclear energy and aerospace. Direct technological benefits accrued to the U.S. in the two decades after 1945 from the legacy of research at wartime laboratories located at MIT, including the development of SAGE, a digital computer and radar system employed to defend the U.S. against possible Soviet missile attacks. As Hughes comments: 'The Whirlwind digital computer developed by Jay Forrester and his MIT colleagues for SAGE proved a seminal device that opened the doors to military and commercial development of interactive, stored-memory digital computers' (Hughes, 1994, p. 427). Furthermore, IBM developed much of its commercial computer hardware, utilizing the learning experience of working on the Whirlwind project. From 1968 until 1972, the DoD were responsible for the development and deployment of a national U.S. computer network that formed the basis of what was to become the Internet, (Smith, 1987, p. 8).

Aerospace Industrial Policy and the U.S. LCA Sector

Because of the USA's staunch adherence to laissez-faire principles, it is assumed that the federal government has not instituted industrial policy in the manner of Europe or Japan. This may be the case, but what has been characteristic of the American economy is a de-facto industrial policy created on the back of military procurement and R&D. As former U.S. presidential economic advisor Laura Tyson asserts, 'The historical record indicates that the United States has had a makeshift, unintentional, but nonetheless effective industrial policy towards its aircraft industry', (Tyson, 1992, p.169).

After 1945, the massive, government-driven expansion of the U.S. aircraft and aero-engine industry, which commenced during the war was extended by an unrivalled government commitment to R&D in high-performance aircraft, both military and civilian. This degree of government support inevitably generated a competitive leading edge in the burgeoning global market for commercial aircraft and the world's airlines were compelled to purchase aircraft, such as the Boeing 707 and 747 if they wished to remain competitive. Similarly, in the military aircraft market, the dominance of the U.S. was maintained through such high-performance aircraft as the F-4, F-5, F-15, F-16, F-104, F-111 and heavy lift transports, including the C-5A and C-130.

This is especially significant since 'it is primarily the steady stream of financing provided by military contracts, which helps to fund the long and risky development of civil products', (Thornton, 1995, p. 27). Furthermore, 'the Federal Government traditionally has funded the lion's share of aerospace R&D, and this support has made US aerospace companies the world's most advanced', (U.S. Senate Democratic Policy Committee).

The U.S. aerospace industry, in particular, has gained much from military/civil technology synergies where, often, whole systems developed for the military were "spun-off" to civil applications, reducing costs and risks for commercial users. The most spectacular case example here is the Boeing 707, which established that company's dominance of the civil jetliner market. Hardy notes the following advantages that accrued to Boeing:

> Without the huge KC-135A programme there would almost certainly have been no Model 707, as its unit costs would have been too high, especially without the benefits of using some KC-135 jigs and tooling... and it was not until 1963, when just over 1000 of the 707, 720, and KC-135/C-135 series had been sold, that Boeing finally passed the break-even point on its jet transport programme, (Hardy, 1982, p. 66).

Regarding the B-707, Hardy's assessment of the value of the military input may be conservative. According to analysis provided at an MIT symposium by March, the U.S. government paid $2bn of the overall Research Development and Production costs, while Boeing contributed a mere $180mn. March's findings are also backed up by Mowery and Rosenberg and Rae, who conclude that Boeing made only a very minor financial contribution to the program, (March, 1989; Mowery and Rosenberg, 1982 and Rae, 1968).

In many instances, products and technologies designed for commercial application have also been able to achieve "spin-on", that is, higher and longer production runs due to the procurement of large military orders, reducing commercial costs and enhancing competitiveness. Commercial gains here have frequently been at the sub-systems level in materials or in manufacturing process technology, broadening the scope of the benefits to be derived from military-funded research and development expenditure.

By participating in the vast U.S. military aerospace research and development process, U.S. aircraft manufacturers are able to pursue technology transfer between military and civil applications in three specific ways.

1.   First, in some cases, direct aircraft-to-aircraft transfer is possible. In such cases, the U.S. government will effectively fund the development of a new military aircraft, allowing the companies engaged on the development process to utilize their new-found expertise (and possibly the aircraft design itself) for commercial purposes. Examples of such direct transfer include the Boeing 707 (from the KC-135) and the Boeing 747 (from the C-5A). In the case of the B-747 the design teams that worked on the original USAF proposal for the transporter moved straight across to the B-747 program when Boeing lost the contract for the C5A to Lockheed, (Newhouse, 1982, p. 113). A more recent case of the same process is the C-17, which now exists in a civil freighter version as the MD-17.

2.   Secondly, U.S. commercial aircraft manufacturers stand to gain as a result of major component transfers between aircraft, particularly aero-engines, most of which in the U.S. were originally designed and developed for use in military jets.

3.   Thirdly, what may be termed minor component transfers are important, especially those which involve aerodynamics, avionics, and specific systems such as navigation.

Under the evolving dual-use policy of the U.S. Department of Defense, developed at the initiative of the Secretary of Defense, such transfers are actively encouraged. The dual-use policy of the U.S. aims to expand the usefulness of DoD R&D spending and to deliver leading-edge, high quality military aircraft. The policy is explicitly conducted with the view of enhancing U.S. technological edge in commercial aircraft markets. Indeed, acknowledging that: 'there are important spin-off economic benefits to civilian technology from these dual-use technologies' (Millburn, 1989,) and that 'we are finding that technology transfer of the dual use technology is going so wonderfully well into the commercial sector' (Fields, 1990), there can be little doubt that the U.S. commercial aircraft industry exists in a highly beneficial symbiotic state with the DoD funded military aircraft sector.

This symbiosis between military and civil aircraft development in the U.S. has been crucial in both fuelling and sustaining its leading edge in technology development. In avionics, for example, military technology continues to migrate to civil aircraft design and production including, for example, data and signal processors, data buses, software elements such as operating systems, and sensors, including infra-red and millimetre wave imagers.

More recently, a development of the greatest potential strategic and commercial significance, has been the development of satellite technology in the form of the U.S. NAVSTAR Global Positioning System (GPS). Originally developed by the DoD in the 1980s to enhance the accuracy of strategic missiles via terminal guidance, the commercial impact of GPS has been described as 'exceeding anything envisioned by the US military' with 'civil applications moving forward at breakneck speed' (Office of Technology Assessment).

The direct and indirect involvement by governments in the aerospace industry offers a good example of how intervention can provide an important stimulus to business growth, technological advance and global competitive edge. In the U.S., Department of Defense research and development programs, implemented principally through the Defence Advanced Research Projects Agency (DARPA), frequently led to technological developments that migrate to the civil aerospace sector.

Examples of civil aerospace projects deriving benefit from military aerospace developments in the U.S. are numerous. For example, Boeing has been helped significantly in its design of large composite structures due to involvement in military programs, particularly through its role as sub-contractor to Northrop Grumman on the B-2 "Stealth" bomber program. Boeing was entrusted with the development of the outboard and aft-centre sections of the B-2 using the latest in advanced composites technology. With regard to the successful use of composites on commercial aircraft the benefits of B-2 work have been substantial for Boeing. On the B-2 program Boeing was able to fabricate the largest composite parts ever made. Boeing's B-2 program manager, Dale Shelhorn, explained the commercial benefits of this in an article in *Aviation Week*:

> Composites are the next generation of aircraft materials, and the B-2 program has been a big boost to composite manufacture…we've made some very important advances in aircraft technology that could be used for commercial aircraft, (*Interview in Aviation Week and Space Technology*, 17 September, 1990, pp-59-62).

In avionics, it has long been recognized that fundamental defence/civil synergies exist. According to a U.S. National Research Council report companies in the supply chain often integrate civil and defence production. In the same report the National Research Council also indicates why the U.S.system guarantees world leadership:

> Frequently, in the case of smaller second- or third-tier suppliers, the military and civil production outputs are sufficiently common that the same facilities and labour pools produce both… Much of the electronics/avionics capability in commercial transports is the by-product of technology developed for military aircraft… In military avionics, the United States still leads the rest of the world; as long as the United States continues this close synergy between civil and military avionics technology, it is doubtful that any foreign country will soon surpass the United States in this technology, (National Research Council, 1985, pp. 100 & 116).

Concerning innovations in avionics, fly-by-light/power-by-wire technology derives from military programs, in particular that of the Sikorsky UH-60/Black Hawk program in 1980. Through NASA's AST program, Boeing received the fly-by-light technology and McDonnell Douglas the power-by-wire elements. The purpose of redirecting these technologies to the civil aerospace sector was to enable U.S. commercial aircraft to access the benefits of full-authority digital computer control. A field in which Airbus Industrie was far ahead of its U.S. rivals at that time.

It is important to note that, unlike much of the European aerospace industry, U.S. aerospace manufacturers are often involved in both military and commercial aircraft development and production simultaneously, allowing at least the potential for seamless technology transfer within the organization. Indeed, at times, the inter-dependency goes even deeper. For example, it is asserted by the Congressional Office of Technology Assessment that:

> During the first 20 years of jet production, the company [Boeing] was carried during that period by steady profits in its military business, especially the B-52 and Minuteman missile, (COTA, 1991, p. 56).

To reinforce this point many examples of technology transfer can be identified in areas such as manufacturing technology, airframe development programs and avionics. For example, in one particular instance work on commercial aircraft (such as the Boeing 737, 757 and 767) and on military aircraft (such as the Black Hawk helicopter and V-22 Osprey Tilt-rotor transport) takes place within the same division and even the same engineering group. In the aerospace supplier company Wyman Gordon's forging and casting division, alloy castings are produced for both commercial and military aircraft by the same employees, the same manufacturing processes and the same equipment.

During the 1980s, the Department of Defense provided the stimulus for the development of advanced technologies and process techniques through its commitment of between $150 and $200 million annually to the ManTech Program. From 1976 through to 1990 funding for this program is estimated to have been $2bn, (Arnold and Porter, 1991, p. 21). Support for sub-contractors in the aerospace industry was also provided by the Department of Defense throughout the decade 1982 to 1992 through the Industrial Modernization and Incentives Program (IMP).

New DoD Policy and Programs

The dramatic changes to defence strategy and operational requirements necessitated by the end of the Cold War encouraged the Department of Defense to radically restructure its procurement policy in 1993. Financial and recoupment incentives were given to U.S. aerospace companies to amalgamate, which led to nearly $100bn worth of merger and acquisition activity in the 1990s. In consequence U.S. companies now dwarf their European counterparts. Following President Clinton's Acquisition Reform proposals, technology research contracts have been placed on a competitive basis, with preference given to those producers dedicated to the pursuit of research applicable to both commercial and military aircraft. The intention here was to initiate a new approach to the fostering of dual-use technology, thereby enabling the Department of Defense to gain access to lower cost, leading edge technology which resides in the civil sector of aerospace. This change in orientation must not be misconstrued. Because the civil industry is now more actively leveraged into defence some industry executives have claimed that the commercial sector now subsidizes defence. But this misses the point of dual-use. Today, because new civil technologies are so fundamental to the U.S.'s defence mission; the civil sector is receiving more direct federal support.

In the U.S., following the end of the Cold War, much of the defence industry has gradually dissolved into a range of high technology industries where global competitiveness is the over-riding goal (Scherpenberg, 1997). As a result, the most promising route to commercial success in the aerospace industry now resides in the greater integration of civil and defence sectors. In its influential report, *Goals for a National Partnership in Aeronautics Research and Technology,* the U.S. Office of Science and Technology Policy illustrates the point perfectly:

> The significant basic technological commonality between military and civil aviation products and services must be exploited to increase the productivity and efficiency of our R&T development activities. This requires government and industry, working together, to actively seek technological goals that are common to both civil and military applications, (OSTP, 1995, p.4).

New over-arching systems integration skills and new integrated production technologies allow civil users enhanced access to leading edge defence research and development while providing the military sector with access to path-breaking civil advances in information technology and microelectronics to drive forward the trend towards information-based warfare.

The Technology Reinvestment Program

A key element in the new Dual-Use Initiative was the Technology Reinvestment Program (TRP), which by 1995 had awarded $800mn to firms seeking DoD funds for dual-use applications, (NSC/OSTP, 1995, p.35) Government statements make transparent the benefits that will accrue to the civil sector: 'As an additional benefit, a dual-use strategy will allow DoD's continuing investments in technology to contribute more to our nation's commercial performance and economic growth', (NSC/OSTP, p. 2).

TRP was implemented by the Defense Advanced Research Projects Agency (DARPA) and the military Services (Air Force, Navy, Army), working jointly with five other agencies: the Department of Commerce (DOC), the Department of Energy (DOE), the Department of Transportation (DOT), the National Science Foundation (NSF) and the National Aeronautics and Space Administration (NASA).

With regard to the commercial aircraft industry a key TRP project was the Advanced Composites for Propulsion Program (ACP). Aiming to reduce the production costs of composites by 30%, the program secured $130mn public funding and was geared to improving commercial competitiveness, (Hearing on Perspectives on the Dual-Use before the Subcommittee on Acquisition and Technology of the Senate Committee on Armed Services, 104th Congress, 1st Session, (1995).

The Technology Reinvestment Program (TRP) set out to achieve technically superior defence systems at reduced cost, while simultaneously strengthening the industrial base on which the Department of Defense depends. The degree to which this funding benefits commercial manufacturers even raised concern in the U.S. Congress in 1995, since the technologies being developed under TRP seemed to be primarily commercial in application with few military spin-offs. Defence funding, therefore, it was alleged, was being diverted through TRP for commercial gain. DARPA's own statements underline the commercial relevance of TRP:

> The United States enjoys a dominant position in the world aircraft market because of its aggressive research and development investment strategy, especially in the Defense sector. However this pre-eminence is being challenged by both Europe and Asia.

> TRP projects in aeronautical technology – a classic dual use area – will provide significant Defense and commercial payoffs in propulsion and engines technologies, flight sensors, aircraft controls structures and design, (DARPA, 1993).

To illustrate the point, two important examples of critical technology transfer acquired by the U.S. commercial aerospace industry, that emanated from RDT&E initially funded by the Department of Defense for military purposes after 1992 are noted below.

Fly-by-Light Advanced Systems Hardware (FLASH)

Although fly-by-light R&T is undertaken at NASA, the main initiative in this field is funded by the Defense Advanced Research Projects Agency (DARPA), through the FLASH program; a $43 million, 24 month program that was the main part of the FY93's $464 million TRP budget.

FLASH was awarded to a team led by McDonnell Douglas (now Boeing) to develop components critical to making fly-by-light flight control technologies viable for military and commercial aircraft. McDonnell Douglas was selected because of its participation in fly-by-wire technology since the 1980s and for its use of fibre optics connections in a limited demonstration on the AV-8B. The company is also part of the NASA fly-by-light/power-by-wire program.

Even without the FLASH program, McDonnell Douglas would have worked on fly-by-light technology anyway, but the TRP funding allowed MDC to accomplish the work twice as fast. Under the DARPA contract, MDC and its partners, including Lockheed Martin and Honeywell, were to retain ownership of the intellectual property rights (IPR).

FLASH exemplifies TRP policy. This program was not intended to develop a military-oriented technology that could eventually transition later to the commercial aircraft sector. FLASH's statement of work already included commercial applications for this technology.

According to DARPA sources, FLASH might save 2,700 kg in aircraft weight, while increasing reliability and maintainability by 10%, it would also reduce wire count by 80%.

FLASH complemented NASA and other DoD fly-by-light development efforts by addressing technical gaps that are critical in making fly-by-light a viable military and commercial technology. FLASH takes a total system design approach to develop all of the technologies required to implement fly-by-light. This approach is divided into three program tasks:

> Task 1 – Fibre Optic Cable Plant : FLASH develops and demonstrates a cost-effective and reliable fibre-optic cable plant that meets future aircraft requirements. Currently, there are few connector, cable, splice, harness, backplane and system test suppliers qualified and certified. FLASH develops capabilities to perform this work as part of the program. This will, in turn, provide viable industry sources of critical fly-by-light technologies.

> Task 2 – FBL Flight Control System : FLASH develops and demonstrates fly-by-light flight control system building blocks, including computers, optical sensors and interfaces, fibre optic-based actuator control loops, and data buses. FLASH also investigates advanced neural network-based control schemes that exploit fibre optics to enable adaptive control algorithms and fault diagnostic systems. This task will serve as the integration task, utilizing components from Tasks 1 and 3 to create complete fly-by-light laboratory demonstration systems for military fixed and rotary-wing aircraft and commercial airliners.

> Task 3 – Actuation: FLASH develops and demonstrates advanced actuation technologies to be integrated into a fly-by-light systems concept. This task involves developing high-power (50 HP) electric actuators, optical servovalves, optical position sensors, and optical interfaces and loop closure.

The overall program's approach is to have researchers define top-level requirements for military fighters, helicopters and transport aircraft, as well as for commercial transport aircraft and helicopters.

Boeing has already tested fly-by-light technology on a MD-11 aircraft, using the U.S. Navy facility at Patuxent River.

Vehicle Management System Integration Technology for Affordable Life Cycle Cost (VITAL)

VITAL is part of a larger program named the 'Affordable Advanced Controls Technologies Program'. MDC (now Boeing) was the winner of the $48 million VITAL competition. With its partners (listed below), MDC (Boeing) developed the components, interfaces, practices, software, and tools needed for affordable vehicle management systems (VMS) that can meet the requirements of both military and commercial aircraft through an open "plug and play" architecture, (DARPA, 1996).

VITAL's goals are to integrate this technology into current aircraft to extend their useful life and set the VMS standard for the next generation of military and commercial aircraft. The DoD intends to develop components for a variety of dual-use applications, hoping therefore to lower the cost of VMS systems. The VITAL project will culminate in flight-tests on military and commercial aircraft to demonstrate its practicality and benefits.

VITAL's team includes many major suppliers to the military and commercial aircraft markets. As for every other TRP program, the team will own the intellectual property rights (IPR). VITAL technology has already transitioned to commercial aircraft programs, such as the Boeing's Active Aeroelastic Wing program. Through this program, a team led by Boeing, and composed of Lockheed Martin Control Systems, and Moog is developing and manufacturing an achievable wing modification incorporating a leading edge aileron and control system. Its ultimate goal is to 'develop control laws to increase roll rate performance p to an estimated 300 degrees per second in the transonic region, while achieving drag reduction and manoeuvre load control without adverse impact on existing structure', (Boeing, 1996).

Tested on a F/A-18 under a joint program between the Air Force's Wright Laboratory and NASA's Dryden, Boeing sources indicated that this program would be used on three commercial programs, the MD-90-40X, the High-Speed Civil Transport, and the Future Thin-Winged Aircraft, (Boeing, 1997).

On a number of occasions in this report we have argued that the assessment of U.S. public subsidy for the LCA sector must include firms in the supply chain. This point is underscored by looking at the companies involved in the VITAL program.

> Aircraft Breaking Systems Corp
> AlliedSignal Aerospace Equipment Systems
> BF Goodrich
> Honeywell Space and Aviation Control
> Lear Astronics Corp
> Litton Guidance and Controls Systems
> Lockheed Martin Control Systems
> Boeing
> Raytheon
> United Technologies.

Small Business Innovative Research

A further way in which the Department of Defense provides support for the U.S. aerospace industry, both military and commercial, is through the DoD's Small Business Innovative Research Program (SBIR) and the Small Business Technology Transfer Research Program (STTR). Since 1995, small businesses have been actively encouraged to target rapid commercialization of their innovative technologies under a Fast Track procedure, involving external funding from private investors. Under this scheme, the Department of Defense can match every $1 received from such investors with up to $4 of Department funds. Schemes such as the SBIR and STTR involve hundreds of small companies and inject several hundred million dollars into the U.S. aerospace industry and its supply chain.

Advanced Technology Program

A supplement to DoD support for U.S. aerospace R&D has come from the Department of Commerce's Advanced Technology Program (ATP), under which industry and government share the costs of project research and development equally and where small businesses and larger enterprises are encouraged to establish joint ventures to develop innovative technology. Estimates suggest that recently Boeing, for example, have received over $50 million annually from ATP support.

Co-operative Research and Development Agreements (CRADAs)

The CRADA program provides a different kind of support for U.S. industry with distinct benefits accruing to aerospace manufacturers. A CRADA is a partnership between a private company and a government facility to research and develop a particular product or innovation through technology transfer to the private commercial sector. Among CRADAs awarded over the last few years are several which provide valuable technological developments for the U.S. aerospace industry, including projects relating to electron beam processing (with Lockheed Martin, Boeing and Northrop Grumman involved in the partnership), enhanced alloys for aircraft parts (involving Boeing); and software development for application on parallel computers (involving Hughes, Olin Aerospace and Boeing, the latter employing these super-computer developments in the HSR program).

Merger and Acquisition Reimbursement

Indirect government support for civil aircraft manufacturers in the U.S. has also become available as a result of the Clinton Administration's drive to achieve a rapid and thorough consolidation and rationalization of the U.S. defence industrial base in recent years. To encourage mergers, merging companies can, under the Morris Trust, pursue their merger with little in the way of tax consequences. Lockheed Martin, for example, is already on record as anticipating corporate savings in excess of $6 billion between 1997 and 1999 and $2.6 billion each year afterwards, (Acquisition and Technology Subcommittee of the Senate Armed Services Committee, April 1997).

It is also important to realize that DoD rules on payment of restructuring costs were changed in the 1990s. In the past DoD always paid a share of the costs of restructuring activities if a single contractor undertook them as part of an internal reorganization. However, DoD did not pay part of the cost if procurement contracts were transferred from one company to another as a result of a business acquisition. However, from July 1993, restructuring costs have been chargeable against contracts that have changed hands because of mergers and acquisitions, (DoD, 1993).

Summary

It is important to note that schemes such as TRP represent only the formal and visible part of the U.S. government's support program for R&D in the U.S. defence and commercial aerospace industry. Additionally, however, there are many other opportunities for U.S. commercial aerospace to derive benefit from the fruits of military-funded R&D work, given the high degree to which these sectors are integrated in the U.S. The analysis here has merely highlighted the more high profile initiatives that have been undertaken.

Quantifying DoD Support of LCA

In the preceding sections we have provided a qualitative analysis of DoD R&D initiatives relevant to the U.S. LCA sector. We now turn to the question of the financial value of DoD benefits to U.S. LCA. Since the quantity of existing DoD contracts (70,000) made it impossible to check for possible LCA-relevant dual-use potential in individual cases, the

101

methodology applied to NASA (i.e. the converging of the "top-down" and "bottom-up" models), could not be applied in our analysis of the DoD. Instead we rely here on the top down model only, which is outlined in Figure 5.1 Our figures for DoD come from the Congressional Budget after it has passed through the Conference Committee stage, (see Figure 4.2 below).

The first assumption in our model of DoD expenditure is that U.S. LCA-relevant dual-use potential exists only in DoD's aircraft activities, rather than in its missiles and space programs. Secondly, the Strategic Defence Initiative (SDI) and Ballistic Missile Defense Office (BMDO) operations, as well as classified programs, have also been excluded from our detailed analysis (see Figure 5.1). Two issues were pertinent here:

➤ With regard to SDI/BMDO, it was assumed that although there may be spin-off from advanced materials R&D, it is difficult to track precisely the relevant civil aircraft related activity.

➤ Although classified programs certainly have a certain LCA-relevance, due to the nature of these contracts, no further details could be extracted with a sufficient degree of precision.

For the purposes of this study the remaining aircraft-relevant DoD activities were further broken down into Research & Technology, Development and Procurement, (see Figure 5.1).

For each of R&T, Development and Procurement, specialists from the European aerospace industry, universities and European aerospace research establishments analyzed and agreed upon the dual-use potential of each individual item of information found in the references listed in Appendix B. Items of information were therefore categorized by these specialists as Purely Civil, Purely Military or Dual-use.

The entirety of these assumptions forms the top-down model for the DoD aerospace-relevant expenditure. The results can be seen below in Tables 5.4 – 5.11 Our analysis shows that both Research & Technology, as well as Development activities, have substantial dual-use potential, while DoD procurement does not reveal any direct dual-use potential, see Table 5.8. The Research & Technology figures include Independent Research and Development funding (IR&D) and the Bids and Proposals (B&P) expenditures from procurement overheads to recoup funding for independent research and development. Hence our figures for LCA benefits include a small percentage of the procurement budget.

IR&D represents costs reimbursed by the DoD when companies undertake independent research and development. IR&D differs from standard R&D in that IR&D projects are undertaken on the independent initiative of the companies, whereas standard R&D is performed in response to specific requests by DoD (Arnold and Porter, 1991, p.15). As a COTA study notes, IR&D is highly significant, 'The IR&D recovery program is important for two reasons. Some corporate executives feel that the dollar values recovered from the government are large enough to be important for a company's overall financial performance. Others feel the real significance of the IR&D program is that the potential for recovery stimulates the company to spend more on R&D', (COTA, 1991, p. 60). According to the study cited above IR&D may be at least 5% of an airframer's overall R&D budget, (COTA, 1991, p.60).

B&P funds are reimbursements to companies for costs incurred in the formulation of bids and proposals for military contracts. These frequently include research and development costs associated with formulating a proposal.

The complete overview of aircraft dual-use potentials for all programs, excluding SDI/BMDO and classified programs, can be seen in Table 5.10. It was assumed in our analysis that dual-use potentials identified in DoD aircraft related Research & Technology activities certainly would form a support to civil Research & Technology activities. It was further presumed that the dual-use potentials identified in DoD aircraft-related development activities would form a considerable support to civil aircraft Research & Technology activities, rather than to civil development activities.

Tracking the benefits that accrue to the U.S. LCA primes from DoD contracts was accomplished by assessing the general information on contract destinations published by the DoD. We estimated that approximately 76% of all DoD contracts for RDT&E and 97% of all DoD contracts for procurement are given to the U.S. aerospace industry. In 1996 the two LCA primes accounted for roughly 21% of all DoD RDT&E contracts to the industry and some 36% of all DoD procurement contracts to the industry, (see Table 5.2).

Table 5.2 **Estimated % Breakdown of DoD Contracts by Destination**

	Average Annual % of Total DoD Contract Volume			
	RDT&E		Procurement (Supplies)	
		%		%
Intramural DoD		17		0
Aerospace Industry-US		76		97
Other		7		3
Total		100		100
	FY 1996 Contract Values U.S. $mn			
	RDT&E		Procurement (Supplies)	
	US$mn	%	US$mn	%
∑ Boeing	1030	9%	1346	5%
∑ MDC	1381	12%	7926	31%
Total LCA Integrating Primes	2411	21%	9272	36%
Other Aerospace Industry-US	9057	79%	16720	64%
Total Aerospace Industry-US	11468	100%	25992	100%

The relative percentage share of contracts of the U.S. LCA Primes, with regard to the total contract volume issued by DoD, was applied to the programs where we had identified clear dual-use potentials. The final result is shown in Table 5.11. Our estimation, based on sound data and robust assumptions, is that a benefit of nearly 600 U.S. $mn p.a. flows from contracts to U.S. LCA Integrating Primes, which have a potential for aircraft dual-use Research & Technology. This is summarized in table 5.3.

Table 5. 3   **Financial Benefits to US LCA Primes from DOD R&T, Development and Procurement, (US$mn)**

	From R&T Contracts		From Development Contracts		From Procurement Contracts (IR&D, B&P)		Total	
Year	Min'm	Max'm	Min'm	Max'm	Min'm	Max'm	Min'm	Max'm
1992	219	260	219	260	126	126	564	645
1993	253	294	257	297	80	80	590	671
1994	236	279	154	197	70	70	459	545
1995	277	319	213	255	66	66	556	640
1996	272	318	263	309	58	58	594	686
1997	279	325	255	301	60	60	593	686

(Due to rounding and selected methodology details do not always add up to totals).

Figure 5.1   Schematic of Flow of DoD R&T Contract Supports to U.S. LCA Primes

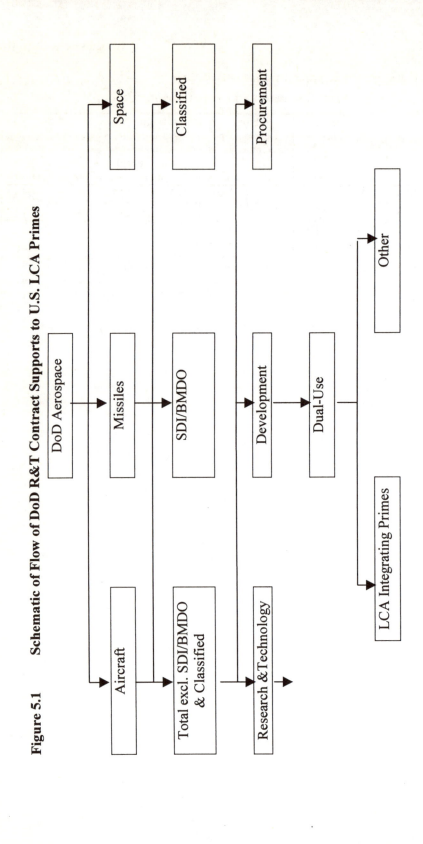

DoD Estimated Expenditure for Aerospace

Table 5.4   DoD Total Aerospace Expenditure (Current US$mn)

	Aircraft		Missiles		Space		Total	
Year	Min	Max	Min	Max	Min	Max	Min	Max
1992	34356	36563	18466	20962	11143	13541	60642	60642
1993	32453	33872	16965	18713	12384	14056	56556	56556
1994	29260	29684	13563	14344	11790	12571	48537	48537
1995	27440	27909	13355	14211	11705	12561	45841	45841
1996	25724	26530	12704	14089	13120	14506	44830	44830
1997	25877	26711	12033	13445	13024	14436	44118	44118

(Due to rounding and selected methodology (see Appendix A) details do not always add up to totals).

Table 5.5   DoD Total Aircraft Expenditure (Current US$mn)

	Total excl. SDI/BMDO, Classified		SDI/BMDO		Classified		Total	
Year	Min	Max	Min	Max	Min	Max	Min	Max
1992	32475	35213	0	0	81	1350	34356	36563
1993	32353	32582	0	0	100	1289	32453	33872
1994	29164	29292	0	0	96	392	29260	29684
1995	27345	27814	0	0	95	95	27440	27909
1996	25724	26530	0	0	0	0	25724	26530
1997	25877	26711	0	0	0	0	25877	26711

(Due to rounding and selected methodology (see Appendix A) details do not always add up to totals).

Table 5.6   **DoD Aircraft excl. SDI/BMDO, Classified (Current US$mn)**

	Research & Technology		Development		Procurement		Total	
Year	Min	Max	Min	Max	Min	Max	Min	Max
1992	3617	5137	5932	6615	24094	24094	34275	35213
1993	3454	4515	6693	7659	21307	21307	32353	32582
1994	3309	4001	6562	7382	18601	18601	29164	29292
1995	3749	4790	6978	7778	15932	15932	27345	27814
1996	3807	5292	7047	7927	14091	14091	25724	26530
1997	3831	5348	6923	7783	14352	14352	25877	26711

(Due to rounding and selected methodology (see Appendix A) details do not always add up to totals).

Table 5.7   **DoD Total Aircraft Research & Technology, incl. IR&D, B&P, excl. SDI/BMDO, Classified Programs (Current US$mn)**

	Purely Civil		Purely Military		Dual-Use		Total	
Year	Min	Max	Min	Max	Min	Max	Min	Max
1992	0	0	1885	3150	1732	1987	3617	5137
1993	0	0	1638	2444	1816	2071	3454	4515
1994	0	0	1634	2055	1676	1946	3309	4001
1995	0	0	1824	2602	1925	2187	3749	4790
1996	0	0	1933	3131	1874	2161	3807	5292
1997	0	0	1911	3140	1919	2208	3831	5348

(Due to rounding and selected methodology (see Appendix A) details do not always add up to totals).

Table 5.8   DoD Total Aircraft Development excl. SDI/BMDO, Classified Programs (Current US$mn)

	Purely Civil		Purely Military		Dual-Use		Total	
Year	Min	Max	Min	Max	Min	Ma	Min	Max
1992	0	0	4561	4988	1372	1627	5932	6615
1993	0	0	5085	5796	1609	1863	6693	7659
1994	0	0	5600	6150	962	1233	6562	7382
1995	0	0	5645	6182	1333	1596	6978	7778
1996	0	0	5396	5990	1650	1937	7047	7927
1997	0	0	5327	5898	1596	1885	6923	7783

(Due to rounding and selected methodology (see Appendix A) details do not always add up to totals).

Table 5.9   DoD Total Aircraft Procurement excl. SDI/BMDO, Classified Programs, IR&D, B&P (Current US$mn)

Year	Civil	Military	Dual-Use	Total
1992	0	24094	0	24094
1993	0	21307	0	21307
1994	0	18601	0	18601
1995	0	15932	0	15932
1996	0	14091	0	14091
1997	0	14352	0	14352

(Due to rounding and selected methodology (see Appendix A) details do not always add up to totals).

Table 5.10   DoD Total Aircraft Dual-Use Research & Technology, excl. SDI/BMDO, Classified Programs (Current US$mn)

	From R&T Contracts		From Development Contracts		From Procurement Contracts (IR&D, B&P)		Total	
Year	Min	Max	Min	Max	Min	Max	Min	Max
1992	1371	1626	1372	1627	361	361	3104	3614
1993	1588	1842	1609	1863	228	228	3425	3934
1994	1476	1747	962	1233	199	199	2638	3179
1995	1736	1998	1333	1596	189	189	3258	3783
1996	1707	1993	1650	1937	167	167	3524	4098
1997	1749	2038	1596	1885	171	171	3515	4093

(Due to rounding and selected methodology (see Appendix A) details do not always add up to totals).

Table 5.11   DoD Total Aircraft Dual-Use Research & Technology, excl. SDI/BMDO, Classified Programs To LCA Integrating Primes (Current US$mn)

	From R&T Contracts		From Development Contracts		From Procurement Contracts (IR&D, B&P)		Total	
Year	Min	Max	Min	Max	Min	Max	Min	Max
1992	219	260	219	260	126	126	564	645
1993	253	294	257	297	80	80	590	671
1994	236	279	154	197	70	70	459	545
1995	277	319	213	255	66	66	556	640
1996	272	318	263	309	58	58	594	686
1997	279	325	255	301	60	60	593	686

(Due to rounding and selected methodology (see Appendix A) details do not always add up to totals).

DoD U.S. LCA Funding: Conclusions

The DoD supports the U.S. large commercial aircraft industry through a variety of policy levers. In the 1990s a number of DoD funded dual-use initiatives have facilitated changes in manufacturing technologies to bring a higher percentage of commercial off-the-shelf (COTS) components and systems into military procurement. These initiatives improve the manufacturing efficiency of the U.S. aerospace industrial base as a whole and give companies such as Boeing substantial benefits. Similarly DoD R&T programs, such as FLASH and VITAL, give direct commercial benefits to the U.S. LCA sector because they are explicitly dual-use in nature.

At the heart of the issue of DoD subsidy of the U.S. LCA sector is the issue of synergy and dual-use. In the public domain the idea that defence/civil synergy has declined to a residual category has been a key assumption of recent debate in the USA. However, we contend that in manufacturing technologies, materials and avionics substantial synergies remain. If they did not the recent U.S. dual-use programs would not make sense. With Boeing now a major defence contractor, having absorbed the defence contracts of MDC, the scope for defence/civil technology transfer must now be substantial for the U.S. LCA sector. Indeed, with the launch of the MD-17 freighter, derived from the C-17 military transport, synergy still exists at the level of the whole airframe.

In order to put a precise figure on DoD LCA financial subsidy we analyzed DoD aircraft expenditure for areas of RTD&E with clear dual-use potential. On the basis of contract volume to LCA primes we then estimated the value of DoD dual-use R&T to Boeing and MDC. On this basis we believe that the U.S. LCA manufacturing companies had a cumulative financial benefit in the period 1992-1997 of U.S. $3,356mn. In other words approximately US$560mn annually for the last six years.

111

6   Federal Subsidy of the U.S. Large Commercial Aircaft Sector: Conclusions

Overview

Our analysis of federal subsidy to the U.S. LCA sector indicates that substantial public funding is provided in the U.S. for LCA. This funding is channelled to the industry chiefly through the Department of Defense (DoD) and the National Aeronautics and Space Administration (NASA).

DoD Subsidy of U.S. LCA

In the case of the DoD, programs are geared to procuring advanced aircraft and systems and to ensuring the competitive strength of the U.S. aerospace/defence industrial base. While the aim of ensuring the global dominance of the U.S. LCA sector may not be the explicit purpose of these programs, this, in effect, is the clear by-product of DoD aerospace initiatives. As the largest budget holder for R&D in the world, the DoD has the power to shape the overall strength of industrial sectors utilized by U.S. defence procurement. Quite simply, to maintain military superiority the U.S. must ensure the health and vitality of the defence/industrial base. Because Boeing, the one remaining U.S. LCA firm, is a major defence contractor it inevitably benefits from DoD R&D, which can be spun-off to civil applications.

Defence/Civil Synergy

In past decades the links between aerospace defence and civil technologies has been obvious. However, in the 1990s U.S. sources have made much of the notion that synergy has declined or even disappeared. Superficially, this seems plausible, as the design characteristics of civil and military aircraft are now so divergent. However, the argument is fundamentally flawed. In

112

materials, avionics and design and manufacturing tools fundamental synergies remain.

In the 1990s the USA has sought to reform defence procurement by merging the defence and civil sectors together. The aim here has been to reduce costs on the defence side by benefiting from commercial best-practice, such as fixed price contracts and lean manufacturing. However, the dual-use strategy that this policy entails would be impossible to implement if civil and military aerospace were really divergent. Moreover, a consequence of the dual-use philosophy has been to target civil industry for more direct financial support, through initiatives such as the Technology Re-investment Program. In our view such programs have heightened the benefits that previously accrued from schemes like ManTech, which, in the 1990s, has been gradually absorbed into the Defense and Manufacturing Science and Technology Program (DMSTP).

Assessing the precise benefits that accrue to the U.S. LCA sector from the extensive number of DoD programs is a daunting task. The major problem is one of visibility and traceability, because the actual transfer of technology takes place inside the premises of U.S. companies. Moreover, because of the sensitivity of the subsidy issue U.S. large commercial aircraft firms are not likely to advertize defence/civil spin-off. However, it is an absolute certainty that this process takes place. Historically, it is well known that Boeing engineers moved easily from defence projects, such as the C-5, to work on the civil B-747 aircraft. On the B-777 Boeing engineers have acknowledged the role that composite manufacture on the B-2 played in supporting the new civil application. Regarding whole aircraft synergy the MD 17 freighter is a direct derivative of the C-17 heavy-lifter.

Because the actual transfer of technology is difficult to quantify, in this report we decided not to put a value on such activities. Instead, we identified areas of DoD programs which we believe had clear spin-off potential and we then estimated the value of the dual-use R&D contracts from these programs going to the U.S. LCA sector. On that basis we contend that a financial benefit of roughly $560mn a year is transferred from DoD programs to the U.S. LCA sector. However, given the overall value of DoD aircraft R&D and the fact that Boeing is now a major defence company this figure is undoubtedly conservative. But we believe it is intellectually defensible. In the past much larger sums have been suggested, but this has not contributed to a sensible debate between the protagonists in the trade dispute.

NASA Subsidy of U.S. LCA

Proving the existence of federal subsidy to U.S. LCA is much easier in the case of NASA than the DoD. As we showed in chapter four a key part of NASA's mission is to ensure the world leadership of U.S. aeronautics. In the 1990s a series of statements by NASA officials have clearly outlined the agency's contribution to U.S. LCA programs. Moreover, the *NASA/DoD Aerospace Knowledge Diffusion Study,* which we have used extensively, makes clear that U.S. aeronautics is a public/private partnership, where significant benefits accrue to U.S. LCA firms from public funding. While the U.S. authorities may continue to contest actual figures, the publication of this DoD/NASA study makes it intellectually impossible to claim that U.S. LCA does not receive financial benefits from the federal government.

As we detailed in chapter four NASA subsidy comes from the Aeronautical Focused Programs and the R&T Base Program. R&T conducted in these programs is transferred to U.S. firms, whose engineers undertake much of the work on NASA R&D contracts. While Boeing officials continue to deny this, NASA documents detail the agency's contribution to the B-777, (see chapter four). In order to make our case cast iron we gave details in the NASA chapter of a specific contract that has already led to the fabrication of a composite wing for current and future applications.

In the case of NASA we were able to supplement our "top-down" data for NASA expenditure with "bottom up" data based on NASA contracts. On the basis of this analysis we estimate that on average roughly $390mn per annum went to U.S. LCA firms in the years 1996 and 1997. However, this excludes contracts to U.S. suppliers, whose systems and components are integrated into LCA. If these companies are included, then U.S. LCA sector benefits can be construed as in excess of $500 million per annum.

Summary

On the basis of our systematic analysis of U.S. data sources and an intensive assessment of R&D contracts we conclude that the U.S. LCA sector received federal financial contributions, or subsidy, (using WTO definitions) in the years of 1996 and 1997 of more than one billion dollars per annum. Such subsidy comes from NASA and the DoD and is underscored by executive policy documents, which emphasize the significance of the LCA industry to the U.S. economy and its need for public support. In short, the subsidy derives from the fact that the USA has a clear industrial policy for its aerospace industry, which indicates a mercantilist orientation to strategic industries

Appendix A: Methodology

Explanation

Data in this study on U.S. federal expenditure for aerospace has been taken from a major empirical investigation of U.S. aerospace expenditure.

Sources

Generally speaking, this study on U.S. government expenditure for aerospace relied on published and/or accessible data only. As the U.S. has clearly democratic structures and since the approval process of any government's budget is an essential task of the legislature in any democracy, it was expected that government publications, which often form the basis for debate and discussion by elected representatives, would prove to be a valuable source of information for the purpose of this analysis. While details of the legislative and budgetary decision process differ between agencies, this analysis showed that a substantial amount of information on aerospace expenditure could indeed be found in these government budgetary publications. Additional information can be found in a variety of documents accompanying the budgetary process, e.g. views on the proposed government budget presented by "rapporteurs" of the legislative bodies.

With regard to classified/confidential data, it must be recognized that much aerospace-relevant data denoted as classified or confidential can be found in U.S. government publications. Although no details are provided, one can find the total amount of expenditure for classified or confidential programs. Knowing where such confidential or classified information is published, i.e. in which chapter, under which headline, etc., there is a certain possibility for categorization. Thus for two reasons, it is not believed that in the United States large amounts of aerospace related expenditures could have been overlooked by our analysis. First of all, such amounts would have to be completely excluded from the budgetary process of a government, which in view of the elected representatives control function over the expenditure is highly improbable in a functioning democracy, and secondly, there are various independent crosscheck possibilities which show very similar figures to the government ones.

Publicly available references also include annual reports of research establishments, space agencies and companies, or even government departments or ministries. In addition, the Internet occasionally served as a valuable source of information. Finally, publicly available information can be found in documents provided by consultancies and analysts. For this analysis, such documents mainly served as backups and sources for crosschecks.

Besides the publicly available information, national aerospace associations provided general support and paved the way for establishing contacts with experts in the government administrations, companies, institutions and agencies. They provided valuable information and data as well as advice for interpreting correctly the data found in the sources. Having compiled some limited amount of data on their own, the national aerospace associations also served as a valuable tool for crosschecking data.

For the United States a number of documents were prepared summarizing the in-depth analysis of the aerospace financing structure of that country, including its specific characteristics. They served as a platform for discussions with specialists from individual research agencies in order to come to a harmonized view. These documents therefore contain considerable expert knowledge about the aerospace funding structures in the United States and the relevant content of aerospace programs.

The analyzed sources generally allowed for the possibility for crosschecking between them. If, for example, the government claims a certain amount to be spent for the domestic aerospace research establishment, then this amount should more or less correspond to the revenue by the same aerospace research establishment from the same national government.

Categorization

The key issue of the methodological approach outlined here is the categorization of each individual item of information found in the various sources. There are various different types of categorization featuring different keywords by which each item of information has been categorized. Below one finds a description of the most important categorizations.

Aerospace Strategic Trade

This methodological approach distinguishes between three different sectors:

AIRCRAFT   Complete systems of and/or airframes for aeroplanes, helicopters and gliders, ground installations, their subsystems and parts, spares and maintenance

Piston engines, turboprops, turbojets, jet engines, their subsystems and parts, spares and maintenance, for installation in aircraft systems

Finished products, subsystems and parts, spares and maintenance, also for test and ground-training equipment, for installation in aircraft systems

MISSILES   Complete systems of and/or airframes for missiles, ground installations, their subsystems and parts, spares and maintenance

Engines, their subsystems and parts, spares and maintenance, for installation in missile systems

Finished products, subsystems and parts, spares and maintenance, also for test and ground-training equipment, for installation in aircraft systems

SPACE   Complete systems of and/or airframes for space vehicles, satellites, launchers, ground installations, …, their subsystems and parts, spares and maintenance

Propulsion devices, their subsystems and parts, spares and maintenance, for installation in space vehicles, satellites, launchers

Finished products, subsystems and parts, spares and maintenance, also for test and ground-training equipment, for installation in space vehicles, satellites, launchers

MSA   Multi-Sectoral Application, if applicable to more than one sector.

118

Categorization by Civil/Military

Items of information were also categorized by *Civil* and *Military*, as well as by *Dual*. While the definitions for *Civil* and *Military* are rather obvious, *Dual* was used whenever research and/or development was identified to be undertaken for dual-use purposes, or if an item of information was believed to represent a spin-off potential between 0% and 100% of its given value.

Categorization by Lifecycle

Each individual item of information has been categorized by the lifecycle segment to which it complies:

Research and Development (R&D), comprising:

➢ All activities in the fields of studies, research and technology as well as demonstrators denoted as Research and Technology activities *(R&T)* including independent R&T financed by overheads in procurement contracts.
➢ *Development* of a product leading to a series production of that product, including acceptance testing and certification tasks. However, in the case of the AEREA research establishments, acceptance testing performed for the Governments was excluded from the definition of Development since this does not represent a technological know-how transfer potential to industry.

Procurement including:

➢ Procurement of OE and spare parts, excl. independent R&T financed by overheads in procurement contracts.
➢ Upgrading
➢ Maintenance (excl. maintenance performed by armed forces themselves).
➢ Operation of existing aerospace hardware. However, this only applies to NASA, mainly for Space Shuttle operations. Expenditures by the armed forces for operating of aerospace equipment (e.g. for aviation fuel) have not been analyzed.

119

Categorization by Structural Flow

Analyses of the various aerospace-funding structures was developed from a generic model which we believe could be applied to all democratic countries. This generic structure can be seen in Figure A1.

Figure A.1 Generic Aerospace Funding Structure

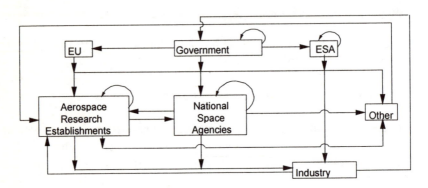

An abstraction of this structure could be achieved by using just four different levels. They are: FROM, VIA 1, VIA 2 and TO. If, for example, a money flow were identified to take place from the government to a national space agency and from there via an aerospace research establishment to the aerospace industry, the four levels would be

- FROM: government
- VIA 1: national space agency
- VIA 2: aerospace research establishment and
- TO: industry.

Figure A2 comprises the chosen abstraction of the expenditure flows over the four levels for each individual key player. ark circles indicate the level at which the total revenues for a given key player can be obtained (e.g. National Space Agency: level 2 "VIA 1"). Level 4 "TO" denotes total intramural expenditure for governments and government organizations, aerospace research establishments, national space agencies and ESA.

120

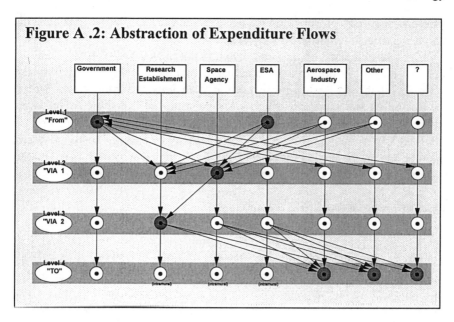

Figure A .2: Abstraction of Expenditure Flows

The complete list of key players included in the various levels can be found in Table A1.

Table A.1 Analyzed Key-Players

Key-Player	Name	Interpretation
AI-US	Aerospace Industry of US	Industry
DOC	Department of Commerce	Government
DOD	Department of Defense	Government
DOE	Department of Energy	Government
FAA	Federal Aviation Administration	Government
GOV	Government	Government
NASA	National Aeronautics and Space Administration	NASA
OTHER	Other	Other

Accuracy

Putting together a database with regard to Government expenditure for aerospace inherently requires accepting some inaccuracies. This is not only true because of the different documents and the different ways in which information is provided in these documents, but also because an analysis of the U.S. immediately involves systematic variations in the ways that

121

different agencies present and categorize data, (see chapter three). This makes it difficult to harmonize all data.

The objective of this statistical analysis was therefore to come as close to reality as possible by creating a model, which is correct in the sense that it should lead to meaningful conclusions and interpretations, but not to figures and numbers which are precise to the last digit. (However, the tables presented in chapters three and four seem to present very precise numbers, but this is entirely due to the necessity of making software-based calculations on the basis of precise data. Each figure presented is estimated to be accurate within 15% and may therefore be rounded up to that extent.)

The Estimated Expenditure Approach

As different sources give different definitions of public funding, e.g. budget, obligation, commitment, burden, outturns, outlays, expenditure etc., which in addition may well be published at different stages of the parliamentary budgetary process (e.g. unvoted proposal, views, voted budget, etc.), there certainly may be an error involved in comparing different items of information from different sources in different countries. Without harmonization, all of these publicly available sources are very difficult to compare.

Our first methodological approach to this fundamental problem was to select, wherever possible, the data closest to real expenditure. This, for example, implied a preference for voted budgets over unvoted budgets, outturns over commitments.

The second methodological approach was the decision to present all results in terms of what we define as *Estimated Expenditure*. This implies that all results represent, to some extent, estimation rather than real and precise expenditure. However, in view of the overall objective of this analysis, this methodological approach was regarded acceptable. The formula for calculating estimated expenditure was as follows:
If, on a national basis,

- $EXP_{curr,\ natcur,\ yr}$ denotes the estimated expenditure of year *yr* measured in current national currency as of year *yr* and
- $D_{yr,\ ref}$ denotes the deflator for year *yr* based on reference year *ref*

then the expenditure of year *yr* measured in constant national currency as of reference year *ref*, $EXP_{const,\ natcur,\ yr,\ ref}$, can be calculated using,

$$EXP_{const,\ natcur,\ yr,\ ref} = EXP_{curr,\ natcur,\ yr}\ /\ D_{yr,\ ref} \cdot [100].$$

The Minimum/Maximum Approach

Looking into very different types of documents in order to extract aerospace relevant items of information, one could not always expect to find the information provided in exactly the way in which it was needed. This implies that most of the items of information, which were found in the documents, could not be precisely categorized.

By definition, a *precise* categorization is given only if a single keyword per type of categorization is sufficient to describe an item of information, e.g. *Aircraft•Civil•R&T*, which clearly contains information on civil aircraft R&T. If more than one keyword is needed in order to categorize the item of information, the latter can not be categorized *precisely*. A typical example is aeronautics R&T where - just from these two words - it cannot be said whether this is R&T for civil or military aircraft or maybe even missiles. Thus, it must be categorized as *Aircraft/Missiles•Civil/Military•R&T*. Our solution to this problem is what we have called the Minimum/Maximum approach.

According to this approach, items of information, which can be precisely categorized, belong to a Minimum set of data. Maximum data then contains – in addition - all items of information, which can not be precisely categorized, i.e. there only is a (mostly unknown) probability that it needs to be taken into account.

Thus, an item of information categorized as *Aircraft•Civil•R&T* would be included under MINIMUM CIVIL AIRCRAFT R&T. MAXIMUM CIVIL AIRCRAFT R&T would – in addition – include an item of information categorized as *Aircraft/Missiles•Civil/Military•R&T*.
Of course, there is a problem of accuracy. The bandwidth between Minimum and Maximum increases with increased "degree of query details". What is meant by "degree of query details" can be explained by the following examples:

- to ask for aircraft R&D data is a more detailed question than to ask for total aircraft data
- to ask for aircraft data is a more detailed question than to ask for total aerospace data.

123

However, even a broad Minimum/Maximum bandwidth correctly reflects the way in which data is presented in the references. It therefore allows the user to get a feeling for the accuracy of the data he or she is looking for.

As a consequence of the selected Minimum/Maximum approach, summing up of data, e.g. *R&D* + *Procurement*, does not necessarily yield the total because:

- Adding up Minimum *R&D* and Minimum *Procurement* does not necessarily result in the Minimum *Total* line because this sum would not take into account those items of information which can only be categorized as *R&D/Procurement*.

- Adding up the Maximum *R&D* line and the Maximum *Procurement* line to arrive at a Maximum *Total* line would not be correct neither, because it would take the items of information, which can only be categorized as *R&D/Procurement*, into account twice, i.e. there is double counting which of course must be omitted.

Therefore the following formulae apply:

$$[\text{Min } Total] \geq [\text{Min } R\&D] + [\text{Min } Procurement]$$

$$[\text{Max } Total] \leq [\text{Max } R\&D] + [\text{Max } Procurement]$$

Of course, the correct data for the total figures is calculated automatically by the used software tool and is also represented in the tables in chapters three and four.

The Dual-Use Approach

Items of information categorized as *Dual* have not only been included in the Maximum figures, but also in the Minimum figures. Minimum *Civil* figures therefore represent purely civil figures as well as dual-use figures, while Minimum *Military* figures include also both, purely military figures as well as dual-use figures. However, when adding up *Civil* and *Military* figures to arrive at *Total* figures, it is ensured that the dual-use items of information are only counted once. In the analysis of United States expenditure the Aerospace Strategy Research Centre assembled two teams of experts to assess civil and military aircraft programs to gauge the degree of dual-use applicability in given program areas. The purpose of this exercise was to make sure any expenditure that was purely military was excluded from assessments of benefits to U.S. LCA.

Multi-Sectoral Application Approach

The Multi-Sectoral Application Approach is very similar to the Dual-Use Approach. However, instead of dealing with *Civil/Military*, the Multi-Sectoral Application Approach deals with the categorization by sectors. If, for example, an item of information (e.g. for generic aerodynamics) is applicable to more than one sector (e.g. aircraft and missiles) this item of information has been categorized by the keyword *MSA* as having a potential for a multi-sectoral application.

Items of information, which have been categorized as providing a potential for the multi-sectoral application, have not only been included under Maximum, but also in the Minimum figures. Thus, figures presented under Minimum Aircraft include all items of information, which are purely aircraft relevant as well as those, which have the potential to be applicable to aircraft and to other sectors, too. The Maximum figures would include, in addition, all those items of information where neither a split by sector could be achieved nor a potential for multi-sectoral application could be identified. Again, when going from individual sectors to Total aerospace figures, multiple counting of figures is eliminated by means of consolidation.

Unknown Destination Approach

While the key-players on each of the four levels of the generic flow structure for most of the analyzed expenditure flows shown in Figures 1 and 2 have been identified, this was not possible for some of the flows.

Minimum data contains all items of information where the destinations of all four levels are known. Maximum data contain – in addition - all items of information where the destination on any of the four levels is unknown.

Summary

The most obvious strength of this methodology is that all sources have been investigated using exactly the same methodology.

The selected Minimum/Maximum approach allows a judgement of the degree of inaccuracy involved in the way the data is provided by the different sources. The margin between Minimum/Maximum therefore allows to be safe, or cautious, depending on the bandwidth size, about any arguments or interpretations which may result from the data.

Since this model has been crosschecked by various groups of people and institutions, it is believed that it draws a reliable picture of the aerospace expenditure and funding structure in USA.

Appendix B: U.S. Sources

100 Companies Receiving the Largest Dollar Volume of Prime Contract Awards, Directorate for Information Operations and Reports, (DoD, annually).

Aerospace Fact & Figures, (Aerospace Industries Association, annually).

Annual Procurement Report, (NASA, annually).

Annual Report to the President and the Congress, (DoD, April annually).

Competitiveness of the European Civil Aircraft Industry, Annex B, Arthur D. Little, (1991).

Defense Science and Technology Strategy, Director, Defense Research and Engineering, (DoD, September 1994).

Defense Technology Area Plan, (DoD, May 1996).

Department of Energy Annual Procurement and Financial Assistance Report, (various years).

European Space Directory, (Sevig Press, various years).

Highlights of the Department of the Navy FY 1998/99 Biennial budget, (Office of Budget, Department of the Navy, February 1997).

Joint Warfighting Science and Technology Plan, (DoD, May 1996).

National Security Science and Technology Strategy, (National Science and Technology Council).

President's BudgetSsubmission, (Ballistic Missile Defence Organization, February annually).

Prime Contract Awards by Service Category and Federal Supply Classification, (Directorate for Information Operations and Reports DoD, annually).

Prime Contract Awards, Directorate for Information Operations and Reports, (DoD, annually).

Procurement Programs (P-1), (DoD, February annually).

Program Acquisition Costs by Weapon System, (DoD February annually).

RDT&E Programs (R-1), (DoD, February annually).

Research, Development, Test and Evaluation (RDT&E R1) Programs, (Forecast International, May 1997).

Second to None: Preserving America's Military Advantage through Dual-Use Technology, (National Economic Council/ National Security Council/Office of Science and Technology Policy, February 1995).

The Military Balance 1997/98, (The International Institute for Strategic Studies, London, October 1997).

United States Space Directory, (Space Publications, various years).

US Government Aerospace Research & Development Contracts, (TDIS. Ltd, Annadale, Virginia, March 1997).

US Government Support of the US Large Civil Aircraft Industry, (Arnold and Porter LLP, Washington, 1991).

Appendix C Congressional Documents

CON-GRESS #	TITLE	FY	PUB-LISHED	REF.	CONGRESSIONAL INFORMATION SERVICE ABSTRACTS OF CONGRESSIONAL PUBLICATIONS		COMMENTS
					#	[Year]	
104	DoD Appropriation Bill 1996	1996	27/07/95	H. Rpr. 104-208	H 183-17	1995	
104	Making Appropriation for the Department of Defense for FY ending 30 September 1996	1996	15/11/95	H. Rpr. 104-344	H 183-30	1995	H 183: Reports Appropriations Committee
103	DoD Appropriation Bill 1995	1995	27/06/94	H. Rpr. 103-562	H 183-18	1994	
103	Making Appropriation for the Department of Defense for FY ending 30 September 1995	1995	26/09/94	H. Rpr. 103-747	H 183-31	1994	
103	DoD Appropriation Bill 1994	1994	22/09/93	H. Rpr. 103-254	H 183-22	1993	
103	Making Appropriation for the Department of Defense for FY	1994	09/11/93	H. Rpr. 103-339	H 183-35	1993	

	ending 30 September 1994						
102	DoD Appropriation Bill 1993	1993	29/06/92	H. Rpr. 102-627	H 183-12	1992	
102	Making Appropriation for the Department of Defense for FY ending 30 September 1993	1993	05/10/92	H. Rpr. 102-1015	H 183-32	1992	
102	DoD Appropriation Bill 1992	1992	04/06/91	H. Rpr. 102-95	H 183-10	1991	
102	Making Appropriation for the Department of Defense for FY ending 30 September 1992	1992	18/11/91	H. Rpr. 102-328	H 183-33	1991	
101	DoD Appropriation Bill 1991	1991	09/10/90	H. Rpr. 101-822	H 183-16	1990	
101	Making Appropriation for the Department of Defense for FY ending 30 September 1991	1991	24/10/90	H. Rpr. 101-938	H 183-27	1990	
101	DoD Appropriation Bill 1990	1990	01/08/89	H. Rpr. 101-208	H 183-16	1989	
101	Making Appropriation for the Department of Defense for FY ending 30 September 1990	1990	13/11/89	H. Rpr. 101-345	H 183-32	1989	
100	DoD Appropriation Bill 1989	1989	10/06/88	H. Rpr. 100-681	H 183-7	1988	
100	Making Appropriation for the Department of Defense for FY ending 30 September 1989	1989	28/09/88	H. Rpr. 100-1002	H 183-28	1988	
100	DoD Appropriation Bill 1988	1988	28/10/87	H. Rpr. 100-410	H 183-18	1987	
100	Making Further Continuing	1988	22/12/87	H. Rpr. 100-498	H 183-21	1987	not used!

	Appropriation for the FY ending 30 September 1988						incl. DoD
99	DoD Appropriation Bill 1987	1987	14/08/86	H. Rpr. 99-793	H 183-19	1986	
99	Making Continuing Appropriation for FY ending 30 September 1987	1987	15/10/86	H. Rpr. 99-1005	H 183-27	1986	not used! incl. DoD
99	DoD Appropriation Bill 1986	1986	24/10/85	H. Rpr. 99-332	H 183-23	1985	
99	Making Further Continuing Appropriation for FY ending 30 September 1986	1986	16/10/85	H. Rpr. 99-443	H 183-32	1985	not used! incl. DoD
98	DoD Appropriation Bill 1985	1985	26/09/84	H. Rpr. 98-1086	H 183-29	1984	
98	Making Continuing Appropriation FY ending 30 September 1985	1985	10/10/84	H. Rpr. 98-1159	H 183-32	1984	not used! incl. DoD

DATA FOR YEAR	WHICH DATA?	TYPE OF DATA	CON-GRESS #	TITLE	FY	PUBLISHED	REF.	CONGRESSIONAL INFORMATION SERVICE ABSTRACTS OF CONGRESSIONAL PUBLICATIONS		COMMENTS
								#	[Year]	
1985	All, except for "Construction of Facilities"	Committee of Conference Authorization	98	Authorizing Appropriations for the National Aeronautics and Space Administration	1985	27/06/84	H. Rpt. 98-873			
1985	Data for "Construction of Facilities"	H. Rpt. Authorization	98	Authorizing Appropriations for the National Aeronautics and Space Administration for FY 1985	1985	21/03/84	H. Rpt. 98-629			
1986	All, except for "Construction of Facilities"	Committee of Conference Budget	99	National Aeronautics and Space Administration Authorization Act, 1986	1986	19/11/85	H. Rpt. 99-379			

1986	Data for "Construction of Facilities"	H. Rpt. Authorization, normated by AEMCA to total of respective data in Committee of Conference Report 99-379	99	Authorizing Appropriations to the National Aeronautics and Space Administration for FY 1986	1986	22/03/85	H. Rpt. 99-32	H7031	1985
1987	All	H. Rpt. Authorization	99	National Aeronautics and Space Administration Authorization Act, 1987	1987	16/09/86	H. Rpt. 99-829	H703-17	1986
1988	All, except for "Construction of Facilities"	"Estimated" (Outlays?)	100	Multiyear National Aeronautics and Space Administration Authorization Act	1989	24/05/88	H. Rpt. 100-650	H703-6	1988
1988	Data for "Construction of Facilities"	H. Rpt. Authorization	100	National Aeronautics and Space Administration Authorization Act, FY 1988	1988	07/07/87	H. Rpt. 100-204	H703-10	1987
1989	All	H. Rpt. Authorization	100	Multiyear National Aeronautics and	1989	24/05/88	H. Rpt. 100-650	H703-6	1988

1990	All, except for "Construction of Facilities"	H. Rpt. Appropriation	101	National Aeronautics and Space Administration Multiyear Authorization Act of 1990	1991	26/09/90	H. Rpt. 101-763	H703-14	1990	
1990	Data for "Construction of Facilities"	H. Rpt. Authorization, normated by AECMA to total respective data in H. Rpt. Authorization Report 99-829	101	National Aeronautics and Space Administration Multiyear Authorization Act of 1989	1990	31/08/89	H. Rpt. 101-226	H703-5	1989	
1991	All, except for details (but not Total) of "Construction of Facilities"	(Outlays?)	102	National Aeronautics and Space Administration Multiyear Authorization Act of 1991	1992	25/04/91	H. Rpt. 102-41	H703-2	1991	
1991	Data for details of			Space Administration Act						Estimated by AECMA by

										means of interpolation
1992	"Construction of Facilities"									
1992	All, except for details (but not Total) of "Construction of Facilities" H. Rpt. Appropriations	H. Rpt. Appropriation	102	National Aeronautics and Space Administration Multiyear Authorization Act of 1992	1993	22/04/92	H. Rpt. 102-500	H703-1	1992	
1992	Data for details of "Construction of Facilities"									Estimated by AECMA by means of interpolation
1993	All, except for details (but not Total) of "Construction of Facilities"	H. Rpt. Appropriation	103	National Aeronautics and Space Administration Authorization Act, FYs 1994 and 1995	1994	10/06/93	H. Rpt. 103-123	H703-2	1993	

1993	Data for details of "Construction of Facilities"									Estimated by AECMA by means of interpolation
1994	All, except for details (but not Total) of "Construction of Facilities"	H. Rpt. Appropriations	103	National Aeronautics and Space Administration Authorization Act, FYs 1995 and 1996	1995	03/08/94	H. Rpt. 103-654	H703-7	1994	
1994	Data for details of "Construction of Facilities"									Estimated by AECMA by means of interpolation
1995	All, except for details (but not Total) of "Construction of Facilities"	H. Rpt. Funding	104	National Aeronautics and Space Administration Authorization Act, FY 1996	1996	04/08/95	H. Rpt. 104-233	H703-8	1995	
1995	Data for									Estimated by

										AECMA by means of interpolation
1996	details of "Construction of Facilities"									
1996	All, except for details (but not Total) of "Construction of Facilities"	Senate Appropriation	104	National Aeronautics and Space Administration Authorization Act, FY 1997	1997	22/07/96	Senate 104-327	S263-34	1996	
1996	Data for details of "Construction of Facilities"	NASA Budget Authority								www.hq.nasa.gov
1985-1996	Data for details (but not Total) of "Research and Program Managem ent"									AECMA estimation based on NASA Civil Service Employee Distribution as published in H. Rpt. 99-32, see sheet NASA_ Employees

Appendix D NASA Contracts

Table D 1 LCA Benfits from NASA Focused Programs

Estimated Average Annual Expenditure

on the basis of years 1996-1997
Subcontracted Work from NASA Aeronautics Aeronautical Focused Programs

Current Mio US$

	to US Aerospace Industry									To Other			Total		
	LCA Integrating Prime(s)			Other			Total								
	LCA relevant	Other	Total	LCA relevant	Other	Total	LCA relevant	Other	Total (1)	LCA relevant	Other	Total (1)	LCA relevant	Other	Total (1)
Purely Civil	164	0	164	2	41	43	166	41	208	0	112	112	166	153	319
Purely Military	/////	NA	NA	/////	NA	NA	/////	33	33	/////	18	18	/////	51	51
Dual-use	40	40	40	12	31	31	52	70	70	41	38	38	93	108	108
Total	204	NA	NA	14	NA	NA	218	145	311	41	167	167	260	312	479

Table D 2 LCA Benefits from NASA the R&T Base Program

Estimated Average Annual Expenditure

on the basis of years 1996-1997

Subcontracted Work from NASA R&T Base Program

Current Mio US$

| | to US Aerospace Industry | | | | | | | | | To Other | | | Total | | |
| | LCA Integrating Prime(s) | | | Other | | | Total | | | | | | | | |
	LCA relevant	Other	Total	LCA relevant	Other	Total	LCA relevant	Other	Total (1)	LCA relevant	Other	Total (1)	LCA relevant	Other	Total (1)
Purely Civil	150	0	150	2	37	39	152	37	189	0	102	102	152	139	291
Purely Military	NA	NA	NA	/////	NA	NA	/////	30	30	/////	16	16	/////	47	47
Dual-use	36	36	40	11	28	28	47	64	64	38	35	35	85	99	99
Total	186	NA	NA	13	NA	NA	199	132	2841	38	153	153	237	285	436

Table D 3 Estimated Expenditure of all NASA R&T Contracts to the LCA and LCA Relevant Sector

Estimated Average Annual Expenditure
on the basis of years 1996-1997
Subcontracted Work from NASA Aeronautics Aeronautical Programs
Current Mio US$

| | to US Aerospace Industry | | | | | | | | | To Other | | | Total | | |
| | LCA Integrating Prime(s) | | | Other | | | Total | | | | | | | | |
	LCA relevant	Other	Total	LCA relevant	Other	Total	LCA relevant	Other	Total (1)	LCA relevant	Other	Total (1)	LCA relevant	Other	Total (1)
Purely Civil	314	0	314	4	78	83	318	78	397	0	214	214	318	292	611
Purely Military	NA	NA	NA	////	NA	NA	////	64	64	////	34	34	////	98	98
Dual-use	76	76	76	23	59	59	99	135	135	79	73	73	178	207	207
Total	390	NA	NA	27	NA	NA	417	276	595	79	320	320	496	597	915

Table D 4 NASA Active Contracts 1996/97

Selection of NASA Aeronautics Contracts still active in 1996 /97

AWARD DATE	RECIPIENT	TITLE	CONTRACT VALUE [US$]	CIVIL/MILITARY	R+TB/AFP
1989	BASF	Advanced Composite Structural Concept	2,416,500	DUAL	R&TB
	Boeing Co.	Advanced Aircraft Propulsion Technology	9,999,992	CIVIL	R&TB
	Boeing Commercial Aircraft Group	Advanced Composite Aircraft Structures	26,499,388	CIVIL	AFP
	Bolt Beranek & Newman	Advanced Aviation Automation	4,826,449	DUAL	R&TB
	Dow Chemical	Novel Material Resins for Primary Structural Composites	4,228,576	DUAL	R&TB
	Grumman Aerospace Corp.	Novel Composites for Wing and Fuselage Applications	3,722,135	DUAL	R&TB
	Lockheed Martin	Advanced Composite Structural Concept	21,050,000	DUAL	R&TB
	Lockheed Martin	Technical Support for R&T	206,354,973	DUAL	R&TB
	McDonnell Douglas	Aircraft & Spacecraft Guidance & Control Technology	4,399,000	DUAL	R&TB
	McDonnell Douglas	ICAPS Phase B	30,236,344	DUAL	AFP
	Northrop Grumman Corp.	Advanced Propulsion/Aircraft System	9,915,170	MILITARY	R&TB
	Research Triangle Institute	Research in Microwave and Antenna Technology	12,830,669	DUAL	R&TB
1990	Georgia Institute Technology	Research in Acoustics and Noise Control	3,952,335	CIVIL	R&TB
	Lockheed Martin	Research in Propulsion / Airframe Integration	8,101,748	DUAL	R&TB
	McDonnell Douglas	Propulsion/Airframe Integration Technology	7,800,000	DUAL	R&TB
	McDonnell Douglas	Research in Acoustics and Noise Control	4,352,524	CIVIL	R&TB
	Research Triangle Institute	Fault-tolerant Integrated Flight System Design and Validation	12,734,140	DUAL	R&TB

Year	Company	Project	Amount	Type	Category
1991	Boeing Co.	Aircraft Composite Primary Structures	10,255,000	CIVIL	AFP
	Boeing Co.	High Speed Research System Studies	19,857,570	CIVIL	AFP
	Lockheed Martin	Aircraft Composite Primary Structure	12,701,000	DUAL	R&TB
	McDonnell Douglas	High Speed Studies	16,538,903	DUAL	AFP
	Northrop Corp.	Aircraft Composite Primary Structure	7,913,048	DUAL	R&TB
	Vigyan	Physical R&D in Flight Control, Guidance and Navigation	3,995,700	DUAL	R&TB
1992	Analytical Services & Mat	Advanced Material Technology	4,910,332	DUAL	R&TB
	Analytical Services & Mat	Basic Research in Aeronautics	4,686,140	DUAL	R&TB
	Bell Helicopter Textron	Composite Primary Structures for Civil Aircraft & Helicopters	1,295,920	CIVIL	AFP
	Lockheed Advanced Dev. Co.	Technologies Applicable to Advanced Aircraft	10,092,242	DUAL	R&TB
	McDonnell Douglas	Advanced Transport Operating System Research	5,445,930	CIVIL	R&TB
	Old Dominion University	Research in Aerodynamics	9,822,000	DUAL	R&TB
	Rockwell International	Advanced Technology Operating Systems Research	4,710,920	DUAL	AFP
	Vigyan	Applied Aeronautics Research and Development	9,697,811	DUAL	R&TB
	Virginia Polytechnic Institute	Research in Aerodynamics	6,511,700	DUAL	R&TB
1993	Advanced Navigation & Position	Advanced Landing System	3,204,100	DUAL	AFP
	Boeing Co.	Acoustics and Noise Control	9,073,750	CIVIL	AFP
	Boeing Co.	High Speed Civil Transport Material Research	35,952,022	CIVIL	AFP
	Boeing Co.	HSR-1 Projects	25,400,000	DUAL	AFP
	Boeing Commercial Aircraft Group	Materials Research for High Sped Civil Transport	20,552,468	CIVIL	AFP
	Boeing Commercial Aircraft Group	Noise Reduction Technology	8,565,600	CIVIL	R&TB

Year	Company	Project	Amount	Category	Type
	Bolt Beranek & Newman	Airplane Acoustics and Noise Control	3,201,200	CIVIL	R&TB
	Lockheed Martin	Airplane Acoustics and Noise Control	7,071,100	CIVIL	R&TB
	Lockheed Martin	High Speed Civil Transport	7,328,200	CIVIL	AFP
	McDonnell Douglas	Acoustics and Noise Control	10,141,250	CIVIL	AFP
	McDonnell Douglas	Power-By-Wire Technology			
	McDonnell Douglas	HSR-1 Projects	23,900,000	DUAL	AFP
1994	Aspen Systems	Graphite Fiber Reinforced composite Manufacturing	500,000	DUAL	R&TB
	AYT Corp.	R&D in Support of Aeronautic Research	1,191,206	DUAL	R&TB
	Boeing Co.	Fly-by-Light Civil Subsonic Transport	12,082,571	CIVIL	AFP
	Boeing Co.	Integrated Wing Design	23,798,750	CIVIL	AFP
	Boeing Co.	Integrated Wing Design, Technology	22,811,250	CIVIL	AFP
	Boeing Co.	Methodologies for Polymer Composites	18,500,000	DUAL	R&TB
	Boeing Co...	Computational Aerosciences Applications	225,425	DUAL	R&TB
	Boeing Commercial Aircraft Group	High Speed Research Airframe Technology	440,000,000	CIVIL	AFP
	Boeing Commercial Aircraft Group	Integrated Wing Design	17,775,000	CIVIL	AFP
	Honeywell	High Speed Research Flight Deck Systems	26,240,000	DUAL	AFP
	Lockheed Fort Worth Co.	Research in Software for Computational Aerosciences	155,689	DUAL	R&TB
	Lockheed Martin	Fast & Accurate Aeroservoelastic Method	723,510	DUAL	R&TB
	McDonnell Douglas	Advanced Parallel Computing / Research in Computational Aeroscience	721,070	DUAL	AFP
	McDonnell Douglas	Blended Wing Body Technology	2,386,387	DUAL	R&TB
	McDonnell Douglas	High Performance Computing / Aircraft Analysis on Advanced Computing Systems	586,693	DUAL	AFP
	McDonnell Douglas	Integrated Wing Design	18,300,000	DUAL	AFP
1995	Aerometrics	Advanced Signal Processors for Icing Applications	150,368	DUAL	AFP

Organization	Project	Amount	Category	Program
AlliedSignal	Simulated Blade Metal Matrix Composite Disk Low	152,375	DUAL	AFP
AYT Corp.	R&D in Support of Aeronautic Research	536,016	DUAL	R&TB
AYT Corp.	Smart Structures Modeling Technology for Active Noise Control	155,755	CIVIL	R&TB
Boeing Co.	Composite Fuselage Structures for Commercial Transport	24,000,000	CIVIL	AFP
Boeing Commercial Aircraft Group	Technology Verification/ Composite Primary Fuselage Structure / Composite Fuselage for Commercial Aircraft	1,816,000	CIVIL	AFP
Central State University	Manufacturing Technology	50,000	DUAL	R&TB
Clark Atlanta University	Manufacturing Technology	119,706	DUAL	R&TB
Honeywell Technology Center	Advanced Air Transportation Technologies	1,929,596	CIVIL	AFP
Lockheed Martin	Advanced ATM Systems Concepts	1,675,051	CIVIL	AFP
M R J	Support for Numerical Aerodynamic Simulation	22,152,108	DUAL	AFP
McDonnell Douglas	ACT Wing Program / Composite Wing for Commercial Aircraft	121,861,556	CIVIL	AFP
McDonnell Douglas	Advanced Aircraft Studies	9,990,000	DUAL	R&TB
McDonnell Douglas	Aircraft & Spacecraft Guidance & Control Technology	6,533,950	DUAL	R&TB
McDonnell Douglas	Dual Fuel Airbreathing Hypersonic Vehicle Design Study	3,207,934	DUAL	AFP
McDonnell Douglas	Improved Processing of Field Level Repairs	841,956	DUAL	AFP
McDonnell Douglas	Intelligent Aircraft Control System	3,134,306	DUAL	AFP
McDonnell Douglas	MD-11 PCA Experiments	49,129	CIVIL	AFP
McDonnell Douglas	Power-by-Wire	28,075,421	DUAL	AFP
Northrop Grumman Corp.	Advanced Aircraft Studies	3,296,682	DUAL	AFP
Odyssey Research Assoc.	Formal Methods for the Design of Flight-Critical Systems	543,509	DUAL	R&TB
U.S. Air Force	Advanced Aircraft Analysis	900,000	MILITARY	R&TB

1996				
Advanced Modular Power Systems	Inexpensive Polymer Precursor Beta	499,661	DUAL	AFP
Advanced Technologies	Reliance Consolidated Models	90,000,000	DUAL	AFP
Advex Corp.	Manufacturing Technology	2,660,086	DUAL	R&TB
Aeroplas Corp International	Toughen Electron Beam Curable Resins	300,000	DUAL	AFP
AlliedSignal	Propulsion and Noise Reduction for Subsonic Aircraft	15,577,786	CIVIL	R&TB
Allison Engine Co.	Critical Propulsion and Noise Reduction Technologies	9,131,000	CIVIL	R&TB
Alpha Star Corp.	Braided Composite Structure Study	599,813	DUAL	AFP
Aspen Systems	Semi-rigid or Rigid High Temperature Isolation	596,379	DUAL	AFP
Beam Technologies	Active Wing - Tool for Design of Active Flow Systems	599,154	DUAL	AFP
Boeing Co.	Aircraft & Spacecraft Guidance Control Technology	10,115,000	DUAL	R&TB
Boeing Co.	High Alpha Research Vehicle	4,199,857	MILITARY	R&TB
Boeing Co.	Technology Verification of Composite Primary Fuselage ures for Commercial Transport Aircraft	2,000,000	CIVIL	R&TB
Boeing Co..	Power Management Control and Distribution	4,030,441	DUAL	AFP
Bouillon Christofferson Sharir	Precision Aircraft Surface Inspection	589,123	DUAL	AFP
Boulder Nonlinear Systems	High-Speed Ferroelectric Liquid Crystal In-fiber lators	584,535	DUAL	AFP
Dynamic Engineering	Reliance Consolidated Models	90,000,000	DUAL	AFP
Expert Systems Applications	Controlled Smart Composite Structure	597,274	DUAL	AFP
General Electric	Critical Propulsion and Noise Reduction Technologies	62,315,800	CIVIL	R&TB
Lockheed Martin	Aerospace Research and Technology	33,795,635	DUAL	R&TB
Materials & ochemical Res.	Composites	598,683	DUAL	R&TB
Materials & ochemical	Processing of C-C Composites	600,000	DUAL	AFP

	Res.				
	McDonnell Douglas	Advanced Aircraft Technology Demonstration	22,028,424	DUAL	R&TB
	Micro Craft Inc.	Manufacturing Technology	1,025,805	DUAL	R&TB
	Sunstrand Corp.	Power Management Control and Distribution	10,150,624	DUAL	AFP
	TRW	Power Management Control and Distribution	5,854,886	DUAL	AFP
	United Technologies Corp.	Critical Propulsion and Noise Reduction Technologies	26,803,518	CIVIL	R&TB
1997	ACTA Inc.	Software for Advanced Aerospace Structures	499,995	DUAL	AFP
	AD Tech Systems Research	Durability Tests on Acoustic Liner CMC Materials Systems	300,000	DUAL	AFP
	Amita	High Performance Ceramic Composites	1,189,460	DUAL	AFP
	Boeing Co.	Advanced Composite Technology	130,000,000	DUAL	AFP
	Boeing Co.	Intelligent Aircraft	3,134,306	DUAL	AFP
	Boeing Co.	Intelligent Damage Active Control System	99,920	DUAL	AFP
	Boeing Commercial Aircraft Group	Advanced Subsonic Transport Noise Reduction Research	5,997,954	CIVIL	AFP
	Lockheed Martin	High Performance Aircraft	19,867,030	DUAL	AFP
	McDonnell Douglas	Advanced Removable Crew Station	69,982	MILITARY	R&TB
	McDonnell Douglas	Studies in Advanced Aircraft Design	9,972,680	DUAL	AFP

Table D 5 Estimated Annual Expenditure of NASA Active Contracts

Selection of NASA Aeronautics Contracts still active in 1996 / 1997

Estimated Expenditure in current US $

TITLE	AGENCY	RECIPIENT	1989	1990	1991	1992	1993	1994	1995	1996	1997	1998	1999	2000	2001	1989-2001
Acoustics and Noise Control	NASA	Boeing Co	-	-	-	-	881,336	2,509,367	2,652,384	1,732,865	891,756	406,043	-	-	-	9,073,750
Acoustics and Noise Control	NASA	McDonnell Douglas	-	-	-	-	783,410	2,300,444	2,725,664	2,067,195	1,251,721	673,068	339,748	-	-	10,141,250
ACT Wing Program / Composite Wing for Commercial Aircraft	NASA	McDonnell Douglas	-	-	-	-	-	-	9,413,782	27,643,105	32,752,730	24,840,295	15,041,210	8,087,875	4,082,559	121,861,556
Active Wing - Tool for Design of Active Flow Systems	NASA	Beam Technologies	-	-	-	-	-	-	-	58,196	165,697	175,141	114,424	58,884	26,812	599,154
Advanced Air Transportation Technologies	NASA	Honeywell Technology Center	-	-	-	-	-	-	187,422	533,634	564,048	348,506	189,638	86,348		1,929,596
Advanced Aircraft Analysis	NASA	U.S. Air Force	-	-	-	-	-	-	87,417	248,897	263,083	171,878	88,451	40,274	-	900,000
Advanced Aircraft Propulsion Technology	NASA	Boeing Co	546,530	1,628,495	2,200,277	1,996,929	1,475,603	974,430	604,062	361,442	212,224					9,999,992
Advanced Aircraft Studies	NASA	McDonnell Douglas	-	-	-	-	-	-	970,331	2,762,758	2,920,217	1,907,846	981,804	447,044	-	9,990,000
Advanced Aircraft Studies	NASA	Northrop Grumman Corp.	-	-	-	-	-	-	320,208	911,705	963,666	629,586	323,994	147,524	-	3,296,682
Advanced Aircraft Technology Demonstration	NASA	McDonnell Douglas	-	-	-	-	-	-	-	6,748,452	11,315,466	3,964,506	-	-	-	22,028,424
Advanced ATM Systems Concepts	NASA	Lockheed Martin	-	-	-	-	-	-	162,698	463,239	489,641	319,894	164,622	74,957	-	1,675,051
Advanced Aviation Automation	NASA	Bolt Beranek & Newman	263,780	785,985	1,061,953	963,808	712,193	470,304	291,548	174,448	102,429	-	-	-	-	4,826,449
Advanced Composite Aircraft Structures	NASA	Boeing Commercial Aircraft Group	1,696,043	5,042,513	6,452,846	5,429,400	3,686,688	2,230,437	1,266,502	694,958						26,499,388
Advanced Composite Structural Concept	NASA	BASF	132,069	393,526	531,697	482,558	356,580	235,471	145,972	87,343	51,284					2,416,500
Advanced Composite Structural Concept	NASA	Lockheed Martin	1,150,447	3,427,985	4,631,588	4,203,538	3,106,148	2,051,176	1,271,552	760,836	446,731					21,050,000
Advanced Composite	NASA	Boeing Co	-	-	-	-	-	-	-	-	16,813,346	44,798,449	39,550,907	20,563,743	8,273,555	130,000,000

Technology	Agency	Company													Total
Advanced Landing System	NASA	Advanced Navigation & Position	-	-	-	-	311,215	886,101	936,603	611,905	314,895	143,381	-	-	3,204,100
Advanced Material Technology	NASA	Analytical Services & Mat	-	-	-	476,942	1,357,964	1,435,359	937,754	482,581	219,733	-	-	-	4,910,332
Advanced Parallel Computing / Research in Computational Aeroscience	NASA	McDonnell Douglas	-	-	-	-	-	93,258	248,483	219,377	114,061	45,891	-	-	721,070
Advanced Propulsion/Aircraft System	NASA	Northrop Grumman Corp.	541,895	1,614,682	2,181,614	1,979,990	1,463,087	966,164	598,938	358,376	210,423	-	-	-	9,915,170
Advanced Removable Crew Station	NASA	McDonnell Douglas	-	-	-	-	-	-	-	-	69,982	-	-	-	69,982
Advanced Signal Processors for Icing Applications	NASA	Aerometrics	-	-	-	-	-	14,605	41,585	43,955	28,717	14,778	6,729	-	150,368
Advanced Subsonic Transport Noise Reduction Research	NASA	Boeing Commercial Aircraft Group	-	-	-	-	-	-	-	-	3,521,935	2,476,019	-	-	5,997,954
Advanced Technology Operating Systems Research	NASA	Rockwell International	-	-	-	457,573	1,302,816	1,377,068	899,671	462,983	210,810	-	-	-	4,710,920
Advanced Transport Operating System Research	NASA	McDonnell Douglas	-	-	-	420,697	1,235,356	1,463,703	1,110,100	672,184	361,443	182,447	-	-	5,445,930
Aerospace Research and Technology	NASA	Lockheed Martin	-	-	-	-	-	-	3,282,579	9,346,261	9,878,937	6,454,141	3,321,391	1,512,326	33,795,635
Aircraft & Spacecraft Guidance & Control Technology	NASA	McDonnell Douglas	281,550	837,076	1,071,197	901,301	612,004	370,261	210,244	115,366	-	-	-	-	4,399,000
Aircraft & Spacecraft Guidance & Control Technology	NASA	McDonnell Douglas	-	-	-	-	-	634,644	1,806,979	1,909,965	1,247,825	642,148	292,389	-	6,533,950
Aircraft & Spacecraft Guidance Control Technology	NASA	Boeing Co	-	-	-	-	-	-	982,473	2,797,327	2,956,756	1,931,718	994,089	452,638	10,115,000
Aircraft Composite Primary Structure	NASA	Northrop Corp.	-	-	-	611,281	1,794,998	2,126,790	1,612,998	976,697	525,184	265,100	-	-	7,913,048
Aircraft Composite Primary Structure	NASA	Lockheed Martin	-	-	-	981,150	2,881,098	3,413,648	2,588,975	1,567,668	842,957	425,504	-	-	12,701,000
Aircraft Composite Primary Structures	NASA	Boeing Co	-	560,467	1,670,023	2,256,386	2,047,852	1,513,232	999,279	619,466	370,659	217,635	-	-	10,255,000
Airplane Acoustics and Noise Control	NASA	Bolt Beranek & Newman	-	-	-	-	310,933	885,299	935,756	611,351	314,610	143,251	-	-	3,201,200
Airplane Acoustics and Noise Control	NASA	Lockheed Martin	-	-	-	-	686,818	1,955,529	2,066,981	1,350,407	694,939	316,426	-	-	7,071,100

Program	Agency	Contractor	Total	1	2	3	4	5	6	7	8	9	10	11
Applied Aeronautics Research and Development	NASA	Vigyan	9,697,811	-	-	-	433,969	953,088	1,852,045	2,834,806	2,681,952	941,951	-	-
Basic Research in Aeronautics	NASA	Analytical Services & Mgt	4,686,140	-	-	209,701	460,548	894,938	1,369,824	1,295,963	455,166	-	-	-
Blended Wing Body Technology	NASA	McDonnell Douglas	2,386,387	-	151,876	377,485	726,029	822,357	308,640	-	-	-	-	-
Braided Composite Structure Study	NASA	Alpha Star Corp.	599,813	26,841	58,949	114,550	175,334	165,880	58,260	-	-	-	-	-
Composite Fuselage Structures for Commercial Transport	NASA	Boeing Co	24,000,000	-	1,073,980	2,358,689	4,583,414	7,015,536	6,637,256	2,331,126	-	-	-	-
Composite Primary Structures for Civil Aircraft & Helicopters	NASA	Bell Helicopter Textron	1,295,920	-	-	-	-	57,991	127,361	247,489	378,816	358,390	125,873	-
Composites	NASA	Materials & Electrochemica l Res.	598,683	26,791	58,838	114,334	175,003	165,567	58,150	-	-	-	-	-
Computational Aerosciences Applications	NASA	Boeing Co	225,425	-	-	10,088	22,154	43,051	65,895	62,342	21,896	-	-	-
Controlled Smart Composite Structure	NASA	Expert Systems Applications	597,274	26,728	58,699	114,065	174,592	165,178	58,013	-	-	-	-	-
Critical Propulsion and Noise Reduction Technologies	NASA	General Electric	62,315,800	-	3,965,948	9,857,278	18,958,819	21,474,240	8,059,516	-	-	-	-	-
Critical Propulsion and Noise Reduction Technologies	NASA	Allison Engine Co	9,131,000	408,604	897,383	1,743,798	2,669,119	2,525,199	886,897	-	-	-	-	-
Critical Propulsion and Noise Reduction Technologies	NASA	United Technologies Corp.	26,803,518	1,199,435	2,634,215	5,118,818	7,835,043	7,412,575	2,603,433	-	-	-	-	-
Dual Fuel Airbreathing Hypersonic Vehicle Design Study	NASA	McDonnell Douglas	3,207,934	-	-	318,782	899,466	1,389,255	600,430	-	-	-	-	-
Durability Tests on Acoustic Liner CMC Materials Systems	NASA	AD Tech Systems Research	300,000	19,093	47,455	91,271	103,381	38,800	-	-	-	-	-	-
Fast & Accurate Aeroservoelastic Method	NASA	Lockheed Martin	723,510	-	-	32,376	71,106	138,173	211,492	200,088	70,275	-	-	-
Fault-tolerant Integrated Flight System Design and Validation	NASA	Research Triangle Institute	12,734,140	-	-	-	333,959	608,611	1,071,825	1,771,618	2,609,069	3,100,881	2,423,153	815,024
Fly-by-Light Civil Subsonic Transport	NASA	Boeing Co	12,082,571	-	404,786	801,913	1,491,336	2,462,915	3,247,433	2,740,813	933,376	-	-	-
Formal Methods for the Design of Flight-Critical Systems	NASA	Odyssey Research Assoc.	543,509	-	24,322	53,415	103,797	158,875	150,309	52,791	-	-	-	-

Program	Org	Company												Total
Graphite Fiber Reinforced composite Manufacturing	NASA	Aspen Systems				48,565	138,276	146,157	95,488	49,139	22,375			500,000
High Alpha Research Vehicle	NASA	Boeing Co						407,933	1,161,480	1,227,677	802,070	412,756	187,940	4,199,857
High Performance Aircraft	NASA	Lockheed Martin						2,569,471	6,846,247	6,044,300	3,142,619	1,264,392	19,867,030
High Performance Ceramic Composites	NASA	Amita						153,837	409,892	361,879	188,152	75,700	1,189,460
High Performance Computing / Aircraft Analysis on Advanced Computing Systems	NASA	McDonnell Douglas			75,879	202,176	178,494	92,805	37,239				586,693
High Speed Civil Transport	NASA	Lockheed Martin		711,790	2,026,631	2,142,135	1,399,507	720,206	327,931				7,328,200
High Speed Civil Transport Material Research	NASA	Boeing Co		3,492,030	9,942,615	10,509,279	6,865,958	3,533,318	1,608,822				35,952,022
High Speed Research Airframe Technology	NASA	Boeing Commercial Aircraft Group			28,161,363	83,726,671	107,144,076	90,150,611	61,214,344	37,034,532	21,029,202	11,539,200	440,000,000
High Speed Research Flight Deck Systems	NASA	Honeywell			2,548,698	7,256,733	7,670,319	5,011,199	2,578,833	1,174,218			26,240,000
High Speed Research System Studies	NASA	Boeing Co	5,491,657	1,928,771	5,804,645	3,792,311	1,951,576	888,609					19,857,570
High Speed Studies	NASA	McDonnell Douglas	3,751,689	1,277,627	4,445,161	3,371,295	2,041,375	1,097,677	554,080				16,538,903
High-Speed Ferroelectric Liquid Crystal In-fiber Modulators	NASA	Boulder Nonlinear Systems					56,776	161,655	170,868	111,632	57,447	26,157	584,535
HSR-1 Projects	NASA	McDonnell Douglas		2,321,413	6,609,600	6,986,304	4,564,316	2,348,861	1,069,505				23,900,000
HSR-1 Projects	NASA	Boeing Co		2,467,109	7,024,429	7,424,775	4,850,780	2,496,279	1,136,628				25,400,000
ICAPS Phase B	NASA	McDonnell Douglas	1,652,509	4,923,977	6,652,840	6,037,987	4,461,689	2,946,322	1,826,464	1,092,869	641,687		30,236,344
Improved Processing of Field Level Repairs	NASA	McDonnell Douglas				157,589	364,625	236,074	83,668				841,956
Inexpensive Polymer Precursor Beta	NASA	Advanced Modular Power System					48,532	138,182	146,058	95,423	49,106	22,359	499,661
Integrated Wing Design	NASA	Boeing Co			2,311,579	6,581,599	6,956,707	4,544,980	2,338,910	1,064,974			23,798,750
Integrated Wing Design	NASA	Boeing Commercial Aircraft Group			1,726,491	4,915,717	5,195,881	3,394,591	1,746,904	795,416			17,775,000
Integrated Wing Design	NASA	McDonnell Douglas			1,777,484	5,060,907	5,349,346	3,494,853	1,798,500	818,909			18,300,000
Integrated Wing Design, Technology	NASA	Boeing Co			2,215,663	6,308,504	6,668,047	4,356,392	2,241,860	1,020,784			22,811,250

Project	Agency	Company														Total
Intelligent Aircraft	NASA	Boeing Co									1,840,431	1,293,875				3,134,306
Intelligent Aircraft Control System	NASA	McDonnell Douglas							586,649	1,357,369	878,822	311,466				3,134,306
Intelligent Damage Active Control System	NASA	Boeing Co									99,920					99,920
Manufacturing Technology	NASA	Clark Atlanta University							11,627	33,105	34,992	22,861	11,765	5,357		119,706
Manufacturing Technology	NASA	Central State University							4,857	13,828	14,616	9,549	4,914	2,237		50,000
Manufacturing Technology	NASA	Advex Corp.								258,375	735,653	777,580	508,011	261,430	119,037	2,660,086
Manufacturing Technology	NASA	Micro Craft Inc.								99,637	283,689	299,857	195,904	100,815	45,904	1,025,805
Materials Research for High Sped Civil Transport	NASA	Boeing Commercial Aircraft Group					3,846,812	8,900,626	5,762,667	2,042,363						20,552,468
MD-11 PCA Experiments	NASA	McDonnell Douglas							28,848	20,281						49,129
Methodologies for Polymer Composites	NASA	Boeing Co						1,796,910	5,116,218	5,407,809	3,533,048	1,818,156	827,859			18,500,000
Noise Reduction Technology	NASA	Boeing Commercial Aircraft Group					831,979	2,366,837	2,503,845	1,635,820	841,816	383,303				8,565,600
Novel Composites for Wing and Fuselage Applications	NASA	Grumman Aerospace Corp.	203,426	606,148	818,974	743,284	549,240	362,696	224,840	134,534	78,993					3,722,135
Novel Material Resins for Primary Structural Composites	NASA	Dow Chemical	231,105	688,622	930,405	844,417	623,971	412,045	255,432	152,839	89,740					4,228,576
Physical R&D in Flight Control, Guidance and Navigation	NASA	Vigyan			308,667	906,386	1,073,924	814,485	493,184	265,192	133,862					3,995,700
Power Management Control and Distribution	NASA	Sunstrand Corp.								985,933	2,807,179	2,967,169	1,938,521	997,590	454,232	10,150,624
Power Management Control and Distribution	NASA	TRW								568,687	1,619,182	1,711,465	1,118,140	575,411	262,001	5,854,886
Power Management Control and Distribution	NASA	Boeing Co								391,478	1,114,628	1,178,154	769,716	396,106	180,359	4,030,441
Power-by-Wire	NASA	McDonnell Douglas							8,600,962	14,421,661	5,052,798					28,075,421
Power-by-Wire Technology	NASA	McDonnell Douglas					7,770,422	22,124,185	23,385,119	15,278,047	7,862,296	3,579,932				80,000,000
Precision Aircraft	NASA	Bullion								57,222	162,923	172,209	112,508	57,898	26,363	589,123

Project	Agency	Company													Total
Surface Inspection	NASA	Christofferson Shair	-	-	-	-	-	-	-	-	-	-	-	-	-	-
Processing of C-C Composites	NASA	Materials & Electrochemical Res.	-	-	-	-	-	-	-	58,278	165,931	175,388	114,585	58,967	26,849	600,000
Propulsion and Noise Reduction for Subsonic Aircraft	NASA	AlliedSignal	-	-	-	-	-	-	-	1,513,075	4,308,073	4,553,605	2,974,977	1,530,965	697,093	15,577,786
Propulsion/Airframe Integration Technology	NASA	McDonnell Douglas	-	499,224	1,484,246	1,598,124	1,899,372	1,085,163	656,521	372,790	204,559	-	-	-	-	7,800,000
R&D in Support of Aeronautic Research	NASA	AYT Corp.	-	-	-	-	-	115,702	329,431	348,206	227,491	117,070	53,305	-	-	1,191,206
R&D in Support of Aeronautic Research	NASA	AYT Corp.	-	-	-	-	-	-	52,063	148,236	156,685	102,366	52,679	23,986	-	536,016
Reliance Consolidated Models	NASA	Advanced Technologies	-	-	-	-	-	-	-	8,741,724	24,889,708	26,308,258	17,187,803	8,845,083	4,027,423	90,000,000
Reliance Consolidated Models	NASA	Dynamic Engineering	-	-	-	-	-	-	-	8,741,724	24,889,708	26,308,258	17,187,803	8,845,083	4,027,423	90,000,000
Research in Acoustics and Noise Control	NASA	McDonnell Douglas	-	336,232	987,328	1,169,828	887,220	537,226	288,874	145,817	-	-	-	-	-	4,352,524
Research in Acoustics and Noise Control	NASA	Georgia Institute Technology	-	252,962	752,081	962,430	809,785	549,863	332,666	188,896	103,652	-	-	-	-	3,952,335
Research in Aerodynamics	NASA	Virginia Polytechnic Institute	-	-	-	1,218,797	2,820,012	1,825,803	647,088	-	-	-	-	-	-	6,511,700
Research in Aerodynamics	NASA	Old Dominion University	-	-	-	954,014	2,716,297	2,871,108	1,875,762	965,293	439,526	-	-	-	-	9,822,000
Research in Microwave and Antenna Technology	NASA	Research Triangle Institute	701,236	2,089,470	2,823,105	2,562,195	1,893,299	1,250,260	775,053	463,755	272,297	-	-	-	-	12,830,669
Research in Airframe Integration / Research in Propulsion / Aerosciences	NASA	Lockheed Martin	-	518,537	1,541,665	1,972,851	1,659,949	1,127,144	681,919	387,212	212,472	-	-	-	-	8,101,748
Research in Software for Computational Aerosciences	NASA	Lockheed Fort Worth Co	-	-	-	-	-	15,122	43,056	45,510	29,733	15,301	6,967	-	-	155,689
Semi-rigid or Rigid High Temperature Isolation	NASA	Aspen Systems	-	-	-	-	-	-	-	57,926	164,930	174,330	113,894	58,611	26,687	596,379
Simulated Blade Metal Matrix Composite Disk Low	NASA	AlliedSignal	-	-	-	-	-	14,800	42,140	44,541	29,100	14,975	6,819	-	-	152,375
Smart Structures Modeling Technology for Active Noise Control	NASA	AYT Corp.	-	-	-	-	-	15,129	43,074	45,529	29,745	15,307	6,970	-	-	155,755
Software for Advanced	NASA	ACTA Inc.	-	-	-	-	-	-	-	-	64,666	172,300	152,117	79,091	31,821	499,995

Program	Agency	Contractor												Total
Aerospace Structures														
Studies in Advanced Aircraft Design	NASA	McDonnell Douglas	-	-	-	-	-	1,289,801	3,436,620	3,034,066	1,577,505	634,689	9,972,680
Support for Numerical Aerodynamic Simulation	NASA	M R J	-	-	-	-	2,151,640	6,126,217	6,475,371	4,230,512	2,177,060	991,288	-	22,152,108
Technical Support for R&T	NASA	Lockheed Martin	11,277,935	33,604,829	45,403,854	41,207,649	30,449,834	20,107,858	12,465,131	7,458,541	4,379,341	-	-	206,354,973
Technologies Applicable to Advanced Aircraft	NASA	Lockheed Advanced Dev. Co	-	-	-	980,262	2,950,103	1,927,372	991,852	451,619	-	-	10,092,242
Technology Verification of Composite Primary Fuselage Structures for Commercial Transport Aircraft	NASA	Boeing Co	-	-	-	-	-	194,261	553,105	584,628	381,951	196,557	89,498	2,000,000
Technology Verification / Composite Primary Fuselage Structure / Composite Fuselage for Commercial Aircraft	NASA	Boeing Commercial Aircraft Group	-	-	-	-	556,335	932,835	326,830	-	-	1,816,000

Appendix E Panel

In order to ascertain the relevant civil, military and dual-use nature of U.S. budget items the following experts from the organizations listed below met as working groups at AECMA in January and February 1999.

Dr Hans-Henrich Altfeld, Head of Policy Research (AECMA).
Snr. Fabrizio Braghini, Management Support (Alenia Aerospazio).
Mr Peter Bruce, Government Affairs Executive (Airbus UK).
Snr Alider Cragnolini (INTA).
Dr Adrian de Graaf, Associate Director (NLR).
Professor Keith Hayward, Head of Research (SBAC)
M. Bernard Jacquot, Director Technology and Industrial Strategy (Aérospatiale Matra).
Professor Philip Lawrence, Director Aerospace Strategy Research Centre (UWE Bristol).
M. Jean Lastennet (SNECMA).
Professor Dieter Scmitt (University of Munich).
Dr David Thornton (Campbell University, USA).
Snr Antonio Vinolo (CASA).
Hr. Wolfgang Wilhelm, Vice President Corporate Development (DaimlerChrysler Aerospace).

Bibliography

AECMA, *Government Expenditure for Aerospace,* (Brussels, 1999).

Aerospace Industries Association of America: Annual Reports, (1991-1998).

Agreement on Subsidies and Countervailing Measures (ASCM), (WTO, 1995).

Agreement Between the European Economic Community and the Government of the United States Concerning the Application of the GATT Agreement on Trade in Large Civil Aircraft on Trade in Large Civil Aircraft, (July, 1992).

Albrecht, U. Lock, P. and Cohen, J. 'The Reluctant Eurofighter Partner' in R. Forsberg, (ed) *The Arms Production Dilemma,* (MIT Press, 1994).

Arnold and Porter, *U.S. Government Support of the U.S. Aircraft Industry,* (Washington, 1991).

Aviation Week and Space Technology, (5/12/88, 17/9/90, 1/3/93, 17/8/94).

Cantor, D. 'Aircraft Production and the U.S. Economy', in J.W. Fischer, (ed) *Airbus Industrie: An Economic and Trade Perspective,* CRS Report to Congress, (Washington, 1992).

Congressional Office of Technology Assessment, *Competing Economies: Government Support of the Large Commercial Aircraft Industries of Japan, Europe and the United States,* (Washington, 1991).

Fields, C. I. (DARPA), *Hearings on National Defense Authorization Act for Fiscal Year 1991,* House Committee on Armed Services, 101st Congress. 2nd session., (March 1, 7, 8, 15 and April 4, 1990).

Financial Times, (September 3, 1998).

Flight International, (2-8, September 1998), p.18.

Galbraith, J. K. 'On the Economic Image of Corporate Enterprise' in R. Nader and M. J. Green, (eds) *Corporate Power in America,* (Grossman, 1973).

Gansler J.S. 'Management of the Defence Industry of the United States', in Jacobson, C. J. (ed.) *Strategic Power USA/USSR*, (Macmillan, 1990).

Gellman Research Associates, *An Economic and Financial Review of Airbus Industrie*, (Jenkintown, 1990).

Golich V. and Pinelli T. 'Who is Managing Knowledge' *International Studies Association* Conference, Minneapolis, (March 23, 1998).

Government Executive (Vols. 1993-1997).

Hansson, A. 'Government Support of the U.S. Aerospace Industry', European *Community Studies Association*, (Seattle, June 1, 1997).

Hardy, M. J. *Boeing*, (Beaufort, 1982).

Hearing on Perspectives on the Dual-Use before the Subcommittee on Acquisition and Technology of the Senate Committee on Armed Services, 104th Congress, 1st Session, (1995).

Heaton, G. R. 'Commercial Technology Development; A New Paradigm of Public Private *Cooperation, Business in the Contemporary World*, (Autumn, Vol 2, No. 1,1989).

Heppenheimer, T.A. *Turbulent Skies*, (Wiley, 1995).

Hughes, T. P. Beyond the Economics of Technology' in O. Grandstrand (ed) *Economics of Technology*, (Elsevier Science, 1994)., p. 426.

Interavia, (May, 1998).

Journal of Metals, (May, 1992).

Lawrence, P. K. & Dowdall P.G. *Strategic Trade in Commercial Class Aircraft: EU versus USA*, (Royal Institute of International Affairs, 1998).

Lawrence, P. K. and Braddon, D. L. *Strategic Issues in European Aerospace*, (Ashgate, 1999).

Lichtenberg, J. 'U.S. Government Subsidies to Private Military R&D Investment: the Defense Department's Independent R&D Policy', National Bureau of Economic Research, (Reprint No. 1415, 1990).

Lopez , V.C and Yeager, L. 'An Aerospace Profile: the Industry's Role in the Economy - the Importance of R&D', Aerospace Industries Association, (Washington, 1987).

March, A. 'The U.S. Commercial Aircraft Industry and its Foreign Competitors;, (MIT Commission on Industrial Productivity, 1989).

Mechanical Engineering, (July 1993).

Millburn, G .P. 'International Technology Transfer: Who is Minding the Store?' *Hearing before the Subcommittee on International Scientific Cooperation*, House Committee on Science, Space and Technology, (101st Congress, 1st Session, July 1989).

Mowery, D. C. and Rosenberg, N. 'The Commercial Aircraft Industry, in Richard Nelson, (ed), *Government and Technical Progress: a Cross-Industry Analysis,* (Pergammon, 1982).

Napier, D. >1997 Year-End Review and Forecast-An Analysis, Aerospace Industries Association, (Washington, 1998).

NASA FY1996 Appropriation Hearing (104th Congress, 1st Session, 1995).

NASA Office of Aeronautics, (FY1996 Budget report).

NASA, Langley Research Centre, *Spin-Off 97,* (NASA, 1997).

National Aeronautical and Space Act, (1958, P.L. 85-568).

National Council on Competitiveness, *Endless Frontier, Limited Resources: U.S. R&D Policy for Competitiveness*, (Washington D.C. April 1996). National Economic Council – National Security Council – Office of Science and Technology Policy, *Second to None: Preserving America's Military Advantage Through Dual-Use Technology*, (Washington, 1995).

National Research Council, *The Competitiveness of the U.S. Civil Aviation Manufacturing Industry: A Study of the Influence of Technology in Determining Industrial Competitive Advantage,* (Washington, National Academic Press, 1985).

National Research Council, *Aeronautical Technologies for the Twenty-First Century,* (Washington, 1992).

National Research Council, *Defense Manufacturing in 2010 and Beyond,* (National Academic Press, 1999).

Newhouse, J. *The Sporty Game*, (Alfred Knopf, 1982).

Office of Science and Technology Policy, *Goals for a National Partnership in Aeronautics Research and Technology,* (Washington, 1995).

Permanent Mission of the United States to the WTO.

Pinelli, T. et. al, *Knowledge Diffusion in the U.S. Aerospace Industry: the NASA/DoD Aerospace Knowledge Diffusion Project*, (Ablex Publishing, 1997).

Rae, J.B. *Climbing to Greatness: The American Aircraft Industry 1920-1960,* (MIT Press, 1968).

Reich, R. *The Work of Nations,* (Alfred Knopf, 1991).

Roland, A. 'The Impact of War Upon Aeronautical Progress: the Experience of NACA', in A. E. Hurley and R.C. Ehrhart, (eds) Air *Power and Warfare,* (U.S. Air Force, Washington, 1978).

Schendel D. (ed), Strategic *Management Journal*-Special Issue, (Winter, 1996).

Smith, M. R. (ed) *Military Enterprise and Technological Change: Perspectives on the American Experience*, (MIT Press, 1987).

Thornton, D. W. *Airbus Industrie: The Politics of an International Industrial Collaboration,* (Macmillan, 1995).

Tyson, L. A. *Who's Bashing Whom: Trade Conflict in High Technology Industries,* (Institute for International Economics, 1992).

U.S. Congress, Report of the Senate Committee on Commerce, Science and Transportation on the NASA Authorization Act of 1988, (June 24, 1987).

U.S. GAO, Results of Fiscal Year 1997 Audit, (April, 1, 1998).

United States Congressional Budget Office, *Reducing the Deficit: Spending and Revenue Options, 152* (February, 1995).

van Scherpenberg, J. 'Transatlantic Competition and European Defence industries: A New Look at the Trade-Defence Linkage', (*International Affairs,* Vol 73, 1, 1997).

Van Tulder R. and Junne, G. *European Multinationals in Core Technologies,* (Wiley, 1989).